'CHERRY' INGRAM

The Englishman Who Saved Japan's Blossoms

NAOKO ABE

Chatto & Windus

LONDON

1 3 5 7 9 10 8 6 4 2

Chatto & Windus, an imprint of Vintage,
20 Vauxhall Bridge Road,
London SW1V 2SA

Chatto & Windus is part of the Penguin Random House
group of companies whose addresses can be found at
global.penguinrandomhouse.com

 Penguin
Random House
UK

Copyright © Naoko Abe 2019

First published by Chatto & Windus in 2019

penguin.co.uk/vintage

A CIP catalogue record for this book is available
from the British Library

ISBN 9781784742027

Text design by Lindsay Nash
Typeset in 11.2/16 pt Carre Noir Std
by Integra Software Services Pvt. Ltd, Pondicherry

Printed and bound in Great Britain by Clays Ltd, Elcograf S.p.A.

Penguin
for ou
made

'CHERRY' INGRAM

Collingwood Ingram, known as 'Cherry' for his defining obsession, was born in 1880 and lived until he was a hundred, witnessing a fraught century of conflict and change.

After visiting Japan in 1902 and 1907 and discovering two magnificent cherry trees in the garden of his family home in Kent in 1919, Ingram fell in love with cherry blossoms, or *sakura*, and dedicated much of his life to their cultivation and preservation.

On a 1926 trip to Japan to search for new specimens, Ingram was shocked to see the loss of local cherry diversity, driven by modernisation, neglect and a dangerous and creeping ideology. A cloned cherry, the *Somei-yoshino*, was taking over the landscape and becoming the symbol of Japan's expansionist ambitions.

The most striking absence from the Japanese cherry scene, for Ingram, was that of *Taihaku*, a brilliant 'great white' cherry tree. A proud example of this tree grew in his English garden and he swore to return it to its native home. Multiple attempts to send *Taihaku* scions back to Japan ended in failure, but Ingram persisted.

Over decades, Ingram became one of the world's leading cherry experts and shared the joy of *sakura* both nationally and internationally. Every spring we enjoy his legacy. *'Cherry' Ingram* is a portrait of this little-known Englishman, a story of Britain and Japan in the twentieth century and an exploration of the delicate blossoms whose beauty is admired around the world.

To my father
Hiroyoshi Abe
1931–2019

Let me die
Underneath the blossoms
In the spring
Around the day
Of the full moon

Saigyō, 1118–1190

Contents

List of Illustrations

Unless otherwise mentioned, all images are reproduced by kind permission of Ernest and Veryan Pollard

Integrated images

First plate section

Second plate section

All photos by Collingwood Ingram from his 1926 trip to Japan

1 A temple courtyard in Kyoto, tree described as *Prunus subhirtella* 'Autumnalis'
2 '*Yoshino* cherry in Uji', thought to be *Somei-yoshino*
3 At a temple gate in Ishiyama, Shiga Prefecture. The tree is described as *Prunus mumé* (plum tree)
4 Kiyomizu temple in Kyoto
5 *Fugenzō* at Nikko
6 An encounter in Ishiyama, Shiga Prefecture
7 Weeping cherry tree at Daigoji temple in Kyoto
8 In the Yoshino mountains

Third plate section

1 Yokohama Nursery Catalogue from 1926–27 (Photo © The Yokohama Nursery Co. Ltd. / RHS Lindley Collections)
2 *Umineko*, London, 2015 (Courtesy of the author)
3 *Kursar* in Chris Lane's nursery, 2015 (Courtesy of the author)
4 *Kanzan* in *Sakura Zuhu* by Manabu Miyoshi, 1921 (Photo © The Trustees of the British Museum)
5 The Grange, 2015 (Courtesy of the author)
6 Ingram aged 99 at The Grange, 1980
7 *Taihaku* at The Alnwick Garden, 2016 (Courtesy of Margaret Whittaker)
8 *Matsumae* varieties in a private nursery at Windsor Great Park, 2015 (Courtesy of the author)

Prologue

A stone's throw from the western moat of the Imperial Palace in Tokyo, a future King of England thrust a shiny new shovel into the cold, wet soil. Peering at the thin trunk of the young cherry tree he had just planted in the British Embassy compound, Prince William smiled at his entourage.

The tree-planting ceremony in late February 2015 was just another ritual for the 32-year-old prince on his first visit to Japan, hours after meeting Emperor Akihito and Empress Michiko in their sequestered quarters within the moat. Unusually, the star of this occasion was the tree itself.

This 10-foot-tall plant was no ordinary variety of cherry tree. *Taihaku*, or 'Great White', was a rare and spectacular tree, lauded by purists for its large, single white flowers. At one time *Taihaku* became extinct in Japan. And its unlikely revival, in a nation where cherry trees are an omnipresent and enduring symbol, was due to one man – an Englishman no less, Collingwood 'Cherry' Ingram.

Introduction

Every major step of my life started with cherry blossoms, as it does for most Japanese people. Japan is a nation where, contrary to Western habit, many significant beginnings occur in April, when the school and government years start and companies welcome their new recruits. When I was beginning nursery school in the central-Honshu city of Nagoya in April 1962, a friend took a black-and-white photograph of me with my mother, Akiko, underneath the gossamer-thin pink petals of a single cherry tree near the school gates. Everyone did the same thing – everyone. Not to have taken a photograph would have been almost sacrilegious. In the picture I'm clinging onto my mama's arm, anxious about the day ahead, but comforted by the umbrella of flowers overhead.

My father, Hiroyoshi, wasn't in the photo. As a journalist, he was working, always working, writing stories about the corporate businessmen driving Japan's re-emergence as a post-war industrial power.

In 1964 the newspaper for which my father worked transferred him to Tokyo and we left our wooden, *tatami*-floored home in Nagoya aboard one of the first bullet trains to the nation's capital. Tokyo was about to host its first Olympic Games – the nation's proudest moment in decades. It was proof that Japan was back on its feet after the humiliating devastation of defeat and nuclear destruction. At the Takamatsu primary school where I enrolled that spring, Mama and I once again stood under the cherry blossoms at the school entrance for the obligatory photograph.

Junior high school. High school. University. For us, it's always the same: April represents a fresh start, another step in life. The cherry blossoms. The photos. And here I am again, in April 1981, captured under fully-opened blossoms by the family Canon camera on the day that I became a professional journalist.

Japan's attachment to cherry blossoms represents a unique and singular obsession. We're a homogeneous people – 98 per cent of the 127 million population are ethnically Japanese – linked by more than 2,000 years of tradition and a cultural affinity to one plant. Other nations have special flowers, of course. But who could imagine virtually the entire population of Britain or Germany or America visiting parks on one particular weekend to view a flower, no matter how lovely?

At the newspaper where I worked in Tokyo, covering the prime minister's office and later the Ministry of Defence, we would send a young assistant to a nearby park next to the Imperial Palace, laden with plastic sheets and cardboard. There, he would spread these mats under a cherry tree and sit shoeless all afternoon – woe betide anyone wearing shoes on our plastic carpet – to defend our space, in readiness for the evening's *hanami*, or cherry-viewing, party (in Japanese, *hana* means flower and *mi* means seeing). *Hanami* was an annual rite of spring, a communal orgy of cherry-blossom-flavoured rice, pickles, wine, *sake* and sweets, loud singing, corporate bonding and reunions of friends and family.

All my life I had taken the blossoms for granted. What I never considered was why most of the planted trees growing in

Japan — seven out of every ten specimens — were of the same variety, known as *Somei-yoshino*. When I moved to London in 2001, I was puzzled by the varied cherry-blossom landscape throughout the British Isles. The blooms I encountered here were multicoloured — white, pink, reddish, some even greenish — and the trees blossomed at different times, usually from mid-March until mid-May. Some of these trees burst into flower, dropped their petals and then another variety would take over, producing a kaleidoscopic cascade effect of blossom that stretched the cherry season to two months.

In Japan, the season is much more defined. The flowers of each *Somei-yoshino* tree survive for about eight days, no more, and the reason they all blossom together and then lose those blossoms together is that they are clones. And so the *sakura*, or cherry-blossom, culture of the twentieth and twenty-first centuries revolved around the flower's short life and swift, predictable death. The cherry blossom was ephemeral, like life itself.

What on earth had happened, I found myself wondering, to the wild cherries, such as *Yama-zakura*, that grew en masse in the mountains or were planted in the cities during the samurai era in the seventeenth and eighteenth centuries? What happened to varieties cultivated in abundance for hundreds of years by the *daimyō*, the feudal lords who ruled principalities all over Japan, and by cherry aficionados in the ancient city of Kyoto? What, indeed, had happened to the once-prized diversity of the cherry trees, with their wonderfully varied and different flower forms?

Researching a newspaper column about how cherry trees spread in the British Isles, I came across the story of Collingwood Ingram, whose crusade to preserve *Taihaku* and scores of other Japanese cherry-tree varieties is legendary among

Western horticulturalists. To the Japanese and the wider world, it is unknown. As I investigated further, Ingram's name kept popping up, and soon I found myself caught up in a full-scale journey of discovery about the cherry tree, which took me to archives, botanical gardens, horticultural research institutes and temples throughout Japan and the United Kingdom. Along the way the quest became deeply personal, upending views that I had held since birth about a tree I thought I knew intimately.

In the course of my research I read about a visit that the Kent Gardens Association had made in May 2010 to The Grange, a twenty-five-room house in the village of Benenden, which, in 1919, had been bought by Ingram and his wife, Florence. The guest speaker on that visit was Charlotte Molesworth, a topiary specialist, who lived in a cottage next to The Grange with her gardener husband, Donald. Charlotte and Donald knew the

Naoko and her parents, spring 2016

Ingram family well, and she suggested that I should contact Ernest Pollard, Ingram's gentlemanly grandson-in-law. In turn, Pollard invited me to his home near the East Sussex town of Rye. His wife, Veryan, is Ingram's granddaughter.

There, on a round wooden table in a tidy ground-floor office, the Pollards had laid out piles of diaries, sketches, handwritten memos, research papers, books, journals, photographs and newspaper articles. To my joy, I had stumbled across a treasure trove, spanning Collingwood Ingram's 100-year life, from 1880 to 1981. Apart from Ernie, as he insisted I call him, no one had looked at most of this valuable collection. Much of it had sat for years in cardboard boxes, before he started sorting out the material.

Ernie kindly lent me copies of transcribed diaries that Ingram had written during his visits to Japan in 1902, 1907 and 1926. Back home in London, I checked the names of the people Ingram had met on his trips to Japan. They proved a revelation: they included royalty, business leaders and top politicians. They were the cream of Japanese society, key members of a newly emergent industrial power, and all were somehow connected to cherry trees. And there was more. Ingram's notes and diaries contained exquisite descriptions of Japan's natural scenery, and of horticulturalists for whom the trees and their blossoms were far more than a splendid plant – they were treasured institutions.

So I jumped joyfully down the rabbit hole of research. I conducted interviews with Ingram's descendants, his gardener, his housekeeper and others who had known him well. My one-dimensional impressions of Ingram evolved into something far more colourful. There he was, born with a silver spoon in his mouth, the wealthy grandson of the founder of the *Illustrated*

London News. There was Ingram as a sick boy, too weak to attend school, and then as a supremely assured teenager. There was Ingram as a young adventurer in Australia and Japan, when the British Empire was at its peak.

In the world of nature, there was Ingram as an ornithologist, drawing birds in French woods in 1917 and 1918 during the First World War; Ingram as an ecologist, decades ahead of his time, preaching the importance of species diversity to a country, Japan, where conformity tends to prevail; Ingram as an agnostic, arguing religion with his parish priest and extolling Charles Darwin's evolutionary theories.

There was also Ingram as a Second World War patriot, heading Benenden's Home Guard, a unit of older and non-conscripted men preparing for the German invasion of England. Not least, there was Ingram the husband, father, grandfather, colleague and friend, sharing with the world his expertise about the natural environment and, above all, about flowering cherry trees.

Collingwood Ingram was a cherry-tree colossus. A passionate advocate for the blossoms and a leading authority on them, he saved some varieties from extinction. He built the world's biggest collection of cherry-tree varieties outside Japan in his Kent garden. His broader legacy was to spread a diverse cherry-tree culture almost single-handedly across the British Isles and the world at large.

Ingram introduced about fifty different kinds of Japanese cherries to Britain. He was the first person in the world to hybridise cherries artificially. He created his own new varieties. And he named a few existing varieties, whose lineage was unknown. To one cherry he gave the name *Hokusai*, in homage to the world-renowned Japanese painter and printmaker, Katsushika

Hokusai, with whom he shared a great love of Japan's iconic mountain, Mount Fuji. He named another variety *Asano*, after the fallen samurai hero of one of Japan's classic literary texts, *Forty-Seven Rōnin*.

He also wrote a seminal book on cherries and gave away seeds, cuttings and saplings, always for free. Moreover he promoted cherries at every opportunity, with his privileged friends and the public alike. And his favourite flowering cherry? *Taihaku*. 'For quality and size, it stands supreme,' Ingram wrote. He became instrumental in the preservation of *Taihaku*, but how did he get caught up in this mission?

My own research also led me to ponder the historic role of cherry blossoms in Japan over the centuries. Living in England, I had seen a thousand 'Visit Japan' advertisements, often highlighting the same two icons: a snow-capped Mount Fuji and the cherry blossom. Yet I soon found that this harmonious imagery masked far more complex questions about the *sakura* as a national symbol.

In ancient Japan, cherry blossoms had been emblematic of new life and new beginnings. That perception began to change subtly in the second half of the nineteenth century. And it accelerated dramatically in the 1930s, as successive governments used the popularity of *sakura*, and its imperial links, as propaganda tools among an unquestioning people. Rather than focusing on cherry blossom as a symbol of life, the songs, plays and school textbooks now focused more on death. Classic poems were deliberately misinterpreted, and it became the norm to believe that the *Yamato damashii*, or 'true Japanese spirit', involved a willingness to die for the emperor – Japan's living god – much as the cherry petals died after a short but glorious life.

In this political environment, from the late nineteenth century onwards, the newly cultivated *Somei-yoshino* cherries were a convenience. Where once Japanese urban areas had been covered with wild cherries and many different varieties, *Somei-yoshino* clones now predominated and their rapid adoption altered the traditional cherry-blossom landscape. They grew fast – from sapling to maturity in about five years. They were easy to propagate. They were cheap. Most of all, they were beautiful.

When *Somei-yoshino* trees are in full bloom, their delicacy and elegance cloak the nation in a pink mantle. Because they flower before their spring leaves come out, their appearance is in some ways more dramatic than that of many other cherries, which usually flower and produce their first leaves in tandem. Whenever Japan had something to celebrate in the late nineteenth and early twentieth centuries, a single variety alone was planted.

By the late 1880s more than 30 per cent of all cherry trees in Tokyo were *Somei-yoshino*. Millions more were planted nationwide after Japan's military victory against Russia in 1905, and to celebrate Emperor Taishō's accession to the throne in 1912 and Emperor Shōwa's (Hirohito) in 1926. Other cherries were neglected or simply disappeared. Few people cared, and fewer still did anything about it.

My father, who was born in 1931, remembered memorising and repeatedly singing songs about cherry blossoms at his primary school in Okayama Prefecture, as militarism took hold after Japan's invasion of the Chinese province of Manchuria. He also recalled another historical juncture, when the singing of such songs stopped abruptly, after Japan surrendered to the Allies in

August 1945 and the American occupation began. This was a time when photographs of the emperor vanished from school classrooms. My father recalls how he and his classmates were each given a brush and a pot of black ink with which to erase any mention in their textbooks of the emperor's divinity and the so-called 'Greater East Asia Co-Prosperity Sphere', which the Japanese military had previously promoted. Emperor Hirohito declared himself a mere mortal on New Year's Day 1946.

In just a few months after the Second World War ended, the nation's outlook would revolve 180 degrees. Black became white. Enemies became friends. And the single-minded 'cherry ideology' that Japan had pursued for more than half a century would be abandoned in favour of post-war realities.

As I was researching, I found myself considering a speech that Collingwood Ingram had given to some of Japan's most eminent cherry experts and supporters in 1926, in which he warned that many of the flowering tree varieties were in danger of extinction. This was a stark warning about Japan's imminent plunge down a destructive path. But when his pleas fell on deaf ears, this determined Englishman would take it upon himself to save the blossoms.

The more I researched and interviewed, the more stories came to light. I came across several Japanese cherry specialists who risked their lives during the Pacific War (as the Second World War is known) to preserve rare varieties. Then there were the grim experiences of Ingram's daughter-in-law in a Hong Kong prisoner-of-war camp. And years after Ingram's death, 'reconciliation' cherry trees arrived in England, the gift of a Japanese cherry-grower who had grown up near a prisoner-of-war camp in northern Japan.

All this work led to my book, *'Cherry' Ingram: The English Saviour of Japan's Cherry Blossoms*, which was published in Japanese in 2016. I was deeply honoured when it won the Japan Essayist Club Award, a major non-fiction prize in my homeland. Then one day, after giving a speech about the book at the Japan National Press Club in Tokyo, I was asked by some non-Japanese friends whether the book would appear in English and other languages.

Thus began a new life for the Ingram story. For the international edition of the book, more historical and cultural perspective was needed. After all, Japanese people are familiar with concepts such as *hanami* or cherry-blossom viewing, but such Japan-specific experiences needed explaining to Nippon neophytes. I also delved into more documents and conducted more interviews throughout Britain and Japan. At Alnwick Castle, the Duchess of Northumberland showed me the garden where she has planted the world's largest collection of *Taihaku* cherries. In the snowy mountains of Gifu Prefecture in central Japan, an impassioned gardener explained how he and his colleagues were keeping alive a 1,500-year-old tree, the world's second-oldest cherry. And in the far south of Kyushu, itself an island in the west of the Japanese archipelago, the grandson of an innkeeper who had befriended scores of *kamikaze* pilots brought tears to my eyes while describing their final hours. Many of these young men – boys, some of them – had died the day after writing farewell poems to their loved ones that compared their lives to the cherry blossoms.

My goal remained the same: to tell an illuminating tale about the surprising connections that linked one man, one flower and two countries. It was the largely unknown story of Collingwood

Ingram, his long life and uncomplicated philosophy. The story of the cherry blossom, its short life and complex ideology. The story of Britain and Japan, two island nations where decades of peace and friendship were punctuated by a four-year war whose consequences still linger today.

In Japan, Ingram's views about heterogeneity clashed with the nation's homogeneity. Ingram assumed that diversity of views and beliefs, species and varieties was a given, to be encouraged and lauded. A society that embraces difference clashes occasionally, but is robust, energetic and forward-looking. For him, this meant that the more different types of birds and plants – including cherry blossoms – there were, the better.

To Ingram, the way that Japan had lurched into a culture of extreme uniformity was alien, restrictive and potentially dangerous. The disappearance of diversity, highlighted by the extinction of the *Taihaku* cherry, was indicative of Japan's militaristic mood in the 1920s and 1930s. The ubiquity of the lookalike *Somei-yoshino* cherry spoke volumes about the dark path of conformity which the Japanese followed, until their 1945 defeat.

But all this is still to come. The story begins with the young Collingwood Ingram at home in the English countryside, surrounded by his family's madcap menagerie of Japanese Chin dogs and albino sparrows.

Part One

—

The BIRTH of a DREAM

1. Family Ties

Years before the cherry blossoms charmed Collingwood Ingram, there lived a pure albino jackdaw called Darlie. Darlie resided in Collingwood's father's hat, in a cupboard inside the hallway of the family's luxurious eleven-room bungalow in Westgate-on-Sea, an English seaside town. Within the hat, the bird had fashioned a nest using fur pulled from Collingwood's mother's sable cap and bedroom slippers. In the nest, the jackdaw, drawn as she was to shiny objects, had stored a silver pen and some forks.

When a servant rang the gong to announce meals, Darlie flew to the dining room and hopped around the table, helping herself to morsels from each plate. Joining Darlie on these culinary circuits were four albino, or leucistic, sparrows – Isidor, Tiny, Wildie and Zimbi – along with Albine and Bil-Bil, two pink-eyed albino blackbirds that loved scoffing hard-boiled eggs. There were at least a dozen other albino birds in the house, including thrushes, a hedge sparrow, a redpoll, a starling and a swallow.

The genetic mutation that these birds carried left them with poor eyesight, poor hearing and an even poorer chance of finding a mate, and their survival outdoors was not assured. So Collingwood and his mother, Mary, kept the birds indoors, where they lived as part of the family, even travelling with them on overseas trips. When Darlie died, Collingwood and Mary set aside a corner of a cabinet in her memory, in which they placed photographs of her, five of her eggs in cotton wool and a brooch containing her feathers. John Jenner Weir, a friend of Charles Darwin and a significant inspiration to the young Collingwood, would call Darlie 'the most charming bird it has ever been my fate to meet with'.

History doesn't record whether Jenner Weir had any comment about the Ingrams' other compulsion: Japanese Chin dogs. Bred and prized by Japanese nobility and samurai lords, these flat-faced, wide-eyed pets resembled Persian cats in many ways. Having been brought to England after Japan opened its doors to the West in the 1850s, the tiny dogs became exotic fixtures in

Mary Ingram and some of her Japanese Chin dogs

moneyed households throughout Europe. Queen Alexandra, for instance, who had married the future King Edward VII in 1863, had been given a Chin soon after her wedding and had helped to popularise the breed. The Ingrams so loved these dogs that at Westgate-on-Sea, their second home, they kept as many as thirty-five Chins at one time.

Each Chin had distinct variations. Most were black and white, but others were red and white, or gold and black. After dinner, according to Collingwood Ingram's cousin, Edward Stirling Booth, the Chins were 'brought in like a set of children into the drawing room for a short time with two dog nurses in attendance. The dogs used to have very particular habits with regard to meals. Every dog had to be completely indulged. Occasionally one little dog would be rushed out and brought back again, and then another one would be rushed out and brought back again. This was another thing which visitors had to put up with.' Booth also noted the presence in the Ingrams' extensive garden of an African wildebeest.

Even in Victorian Britain, where the foibles of the wealthy were generally overlooked, the Ingrams' collections marked them as atypical. And there was no doubt, among the residents of Westgate-on-Sea, that the Ingrams were unusual. Unusually wealthy, too. The family head was Collingwood's proud father, Sir William James Ingram, the Liberal Party's Member of Parliament for Boston, Lincolnshire. He was also managing director of the *Illustrated London News*, one of Britain's most influential and popular newspapers. Willie, as his friends called him, was an energetic big thinker, much like his father Herbert, the newspaper's founder. Sir William's many critics had other descriptions for him, considering him arrogant, litigious and unforgiving, as indeed they had his father. Further detractors included Sir William's five sisters and his mother, Ann, whose remarriage in 1892 at the age of eighty would plunge the family into open warfare.

Sir William's wife, Mary Eliza Collingwood Ingram, was an Australian whose accent had been smoothed out by elocution lessons in London. The couple, both passionate about birds and the natural world, had met in London and married in November 1874 at Christ Church, Paddington. Their three boys, who called their parents Min and Pids, completed the quintet. The eldest, Herbert or Bertie, and his brother, Bruce, attended an elite boarding school, Winchester College, their father's alma mater.

Collingwood, the baby of the family and a sickly child, had never attended school. So while Bertie studied Virgil's *Aeneid*, Collingwood roamed the countryside, studying birds – wagtails and warblers, whinchats and wrynecks. And while Bruce learned about Whistler's portrait of his mother and Constable's *The Hay Wain*, Collingwood learned to whistle the *whit-whit* call of the quail in the marshes of East Sussex. From his earliest childhood, birds were Collingwood's fixation. At the age of three, his Norwegian nurse had held him over a shrub to look into a hedge sparrow's nest containing a clutch of turquoise-blue eggs. 'The study of birds,' he later recalled, 'and in particular the study of their nests and young became an obsession with me – an obsession that persisted for at least half of my life.'

Nature was the boy's religion, and Darwinism his creed. And one day in 1891, quite by chance, he ran into John Jenner Weir, one of Britain's most accomplished ornithologists and botanists.

That meeting, Ingram recalled, was a transformational, almost evangelical experience: 'The manner in which I came to know that stranger has remained an inexplicable episode in my life.'

> I was only about 10, a shy introspective child who in normal circumstances would have never dreamt of accosting a perfect stranger. Yet that was exactly what I did. I was wandering about the countryside by myself in search of birds, when I saw coming towards me, also alone, an elderly gentleman dressed from head to foot in urban black. He might have been anything — a lawyer, a doctor, a businessman.
>
> There was therefore no ostensible reason why I should have suddenly felt irresistibly drawn towards the man. Was it telepathy or was it intuition? I know not. Anyhow, something seemed to tell me that here at last I had found a kindred spirit. Impelled by an uncontrollable urge, I walked straight up to him, and without so much as a word of explanation, bluntly asked him if he was interested in birds — a fatuous question since I already instinctively knew the answer.

In fact, Jenner Weir kept birds and butterflies in an aviary in his garden in south London, where he experimented to see which variety and colour of caterpillars the birds would eat. Darwin cited a number of Jenner Weir's observations in *The Descent of Man* and other books. For three formative years after they met, Jenner Weir lent Collingwood materials and books about the natural world. He died suddenly in March 1894, aged seventy-one, when his young admirer was just thirteen, but his influence lasted throughout Ingram's life. In his final publication, *Random Thoughts on Bird Life*, self-published when he was ninety-eight years old, Ingram wrote of his 'deepest gratitude for his [Jenner Weir's] encouragement'.

Collingwood was already passionate about collecting all varieties of fauna that interested him. His meetings and correspondence with Jenner Weir further encouraged those pastimes. Diverse species must be protected and preserved: that, to Collingwood, was the norm. Indeed, it was variety that made life so rich and fulfilling for him.

Darwin's theories of evolutionary adaptation through natural selection – the 'survival of the fittest' – which Collingwood discussed with Jenner Weir, argued against the natural survival of the family's albino birds, yet survive they did, at least in small numbers; just as Collingwood himself would defy the odds at his birth and would live for more than 100 years.

2. Mayfair-by-the-Sea

The dark and poisonous fog that enveloped the world's largest city on Monday, 26 January 1880 brought London to a virtual standstill. For three days the pea-souper, as the fog was known, blanketed the English capital, severely limiting visibility. Caused largely by the burning of coal, the toxic mix of sulphur dioxide and combustion particles in the fog killed an estimated 11,000 people. Most of the city's five million residents stayed at home, the 32-year-old William and 29-year-old Mary Ingram among them. While a governess looked after the Ingrams' two young sons, the couple enjoyed some precious time alone in their South Kensington residence. Nine months later, on Saturday, 30 October 1880, Mary gave birth to Collingwood, her third and last child.

For the privileged, London life was engrossing, with theatres and concerts, lectures and the best private clubs, and shopping at Hamleys and Harrods. London was the centre of Queen Victoria's British Empire, replete with well-designed parks, museums and galleries containing priceless sculptures, paintings and treasures from around the globe. The capital's rich were awash with cash and assets. The late nineteenth century was also a golden age of political debate, as William Gladstone, leader of the Liberal Party, and Benjamin Disraeli, the Conservative Party's head, traded barbs in Parliament.

Yet for the majority, life was a struggle. Home to about one in five of the British population, London was dangerous, noisy, smelly and smoky. Although more Londoners were now travelling underground on the Metropolitan and District railways, both of which had started carrying passengers in the 1860s, the city streets were a chaotic mess. The thousands of horses that pulled landaus, buses and hansom cabs were dumping mountains of droppings that attracted disease-carrying flies. To stay warm, the citizens burned coal, which blanketed the city in smoke and soot. Housing conditions for the working class were especially bad, with robbery and violence endemic, particularly in the East End slums. Diarrhoea, whooping cough, smallpox, measles and scarlet fever claimed the lives of thousands each year. Tragically, about one in five babies died in their first year, and the average life expectancy for adult men was just forty-two years.

The Ingrams' riches mattered little when their boy's life hung in the balance. Sick with respiratory diseases from the outset, Collingwood was reared on donkey's milk, which was closer to breast milk than that of cows, goats or sheep. Since ancient Egyptian times the lactose, minerals and proteins in donkey's milk had protected infants from infection and had helped build up their immune system. William and Mary's fear that their son would catch tuberculosis, or an equally deadly disease, meant that he wasn't taken outside much as an infant in London. Indeed, the capital was hardly the place to raise an ailing child.

Mary Ingram had grown up breathing the pure, clean air of her childhood home in South Australia. Born in December 1851, she had delighted in riding Shetland ponies and rearing orphan possums with her seven brothers and sisters on the family's sheep station in Strathalbyn, a small town south-east of Adelaide. Her father, Edward Stirling, originally from near Arbroath in Scotland, had arrived in the Antipodean colony in June 1839 on the maiden voyage of a three-masted barque, *Lady Bute*. Edward, the son of a plantation owner in Jamaica and an African slave, had married Harriett Taylor eight years later and eventually become wealthy in the copper-mining business, before returning with the family to the UK in the late 1860s.

To escape London's pollution, and as the capital's population exploded during the 1870s, William and Mary Ingram sought an out-of-town bolthole. They settled on Westgate-on-Sea, which was becoming known by the capital's elite as 'Mayfair-by-the-Sea'. Two hours by train from London, it was a custom-built resort of large homes on private roads, including Britain's first-ever bungalows. Spotting a business opportunity, William Ingram bought eight houses on a sea-facing terrace in

1878, reserving one for his family, which he called 'Loudwater', after his childhood home.

Nine years later, Ingram bought the largest house in Westgate-on-Sea, called simply The Bungalow, following the deaths of its original owners, Sir Erasmus and Lady Charlotte Wilson. Built on a suitable scale for large Victorian families – and, in this case, their pets – The Bungalow had an expansive polished-pine drawing room, several dining rooms, a conservatory, a veranda, rooms for the Ingrams' six servants, stables and a large underground wine cellar. There was central heating, even in the aviary, where the family kept a pet kea, a carnivorous mountain parrot from New Zealand. (The heating alone wasn't, unfortunately, enough to keep the family's favourite albino swallow alive. The bird was unable to migrate south in winter because of a broken wing. So at 7 a.m. and 10 p.m. each day, from October until March, Collingwood and Mary placed a hot-water bottle on a tray under the bird's cage to keep the temperature at an even sixty degrees Fahrenheit. Despite these efforts, the swallow eventually caught a cold and died.)

During the first decade of his life Collingwood was, as he recalled, a 'puny weakling, prone to bronchial disorders'. However, his health slowly improved, aided by trips to the countryside and the seaside, where doctors extolled the benefits of breathing salt air and bathing in the supposedly 'invigorating' English Channel. Meanwhile, when the family was in London, the boy fell in love with the Natural History Museum, the terracotta-faced tribute to Britain's imperial success that opened in 1881. The museum itself couldn't have been more conveniently situated. The Ingrams' home at 65 Cromwell Road in South Kensington was a two-minute walk away. To

Collingwood, and indeed any Victorian visitor, it was a wonder of nature – a collection of exotic preserved plants and animals from virtually every known country on the planet. Within the museum, one gallery displayed the insect, plant and animal specimens collected by the naturalist Sir Joseph Banks during Captain James Cook's voyage around the Pacific on HMS *Endeavour* from 1768 to 1771. Another section showed the collection that Sir Hans Sloane had made on his visits to the West Indies in the early eighteenth century.

On entering the museum, Collingwood could glance up at the richly coloured, cathedral-like ceiling, where 162 individually painted and gilded panels depicted plants from around the world. Often the young naturalist would find himself a chair, open his sketchbook and spend hours drawing the stuffed birds and other specimens. This quiet contemplation of nature was in sharp contrast to the freewheeling atmosphere within the energetic Ingram household.

3. Triumphs and Tragedies

The Ingrams owed everything to the family patriarch, Herbert, a larger-than-life entrepreneur. Born in 1811, Herbert had made his first fortune in Nottingham in the Midlands, selling 'Parr's Life Pills', a quack medicine that promised 'long life and happiness'. Then, in May 1842, he established the basis for his second fortune when he founded the *Illustrated London News* with two friends, William Little and Nathaniel Cooke. In 1843 Herbert married Ann Little, the eldest sister of his business partner. Over the ensuing twelve years she bore four boys and six girls.

In public, Herbert's story was a typical rags-to-riches narrative: the son of a poor butcher from Boston, in rural Lincolnshire, who became a successful businessman, Member of Parliament and philanthropist. In private, it was a tale full of scandal, for Herbert was a philanderer whose predilection for wine and women cast a shadow over the Ingram family for decades.

In 1851 and again in 1856, Herbert sexually assaulted Emma Goodson, William Little's sister-in-law, according to Herbert's biographer, Isobel Bailey. Emma swore on oath in 1857 that Herbert had entered her bedroom and 'commenced by kissing me and then forced his hand into my bosom'. The 1,000-word oath, which contained other accusations of 'vulgar conduct' and 'disgusting manner', was countersigned by Emma's husband, Charles, her brother William, and William's wife, Elizabeth, presumably in case the alleged assaults became a legal issue. The incident had understandable repercussions. Charles confronted Herbert Ingram, who initially denied the allegations, but eventually apologised. Later that year Ingram took his revenge upon Emma and Charles by ending the lease on the property in which they were living. On New Year's Eve 1857, according to William Little, Ingram visited the offices of the *Illustrated London News* and 'plunged his fist several times into my face with all his might'.

In August 1860 Herbert Ingram and his fifteen-year-old eldest son, also named Herbert, travelled to the United States, in part to escape from his problems at home, as well as to follow the 18-year-old Prince of Wales (later King Edward VII) on a tour of North America. There the two Ingrams both died in

September, along with 300 others, when the paddle steamer on which they were travelling, the *Lady Elgin*, sank in Lake Michigan. Ann, who was forty-eight at the time, inherited all Herbert's assets. The inheritance would later split the Ingram family and lead to multiple lawsuits after Ann married, in April 1892, Sir Edward William Watkin, a railway magnate and first instigator of a Channel tunnel. Neither Collingwood nor his parents or brothers attended his 80-year-old grandmother's nuptials, because Sir William Ingram and his brother were convinced that Sir Edward was marrying his mother solely for her money.

That wasn't the only family drama with which Collingwood grew up. In April 1888, when he was seven, his swashbuckling Uncle Walter, his father's younger brother, was trampled to death by an elephant during a big-game hunt, west of Berbera in Somaliland. A flash flood then swept away Walter's remains.

The irony was that a grisly end had been predicted for Walter, who in 1884 had joined a doomed British relief force sent to Egypt to rescue the besieged General Charles Gordon in Khartoum. Later, after journeying down the Nile, the 32-year-old traveller had bought an Egyptian sarcophagus containing the body of a fourth-century BC Theban priest and had sent it to his brother William at the newspaper. According to H. Rider Haggard and other literary giants of the time, many of whom wrote for the *Illustrated London News*, Walter was cursed for disturbing the priest. The Ingram hex became part of the popular 'curse of the pharaohs' narrative, after the death of Lord Carnarvon following the 1922 discovery of Tutankhamun's tomb in Egypt. Collingwood's brother, Bruce, was a close friend of Lord Carnarvon and of Howard Carter, the tomb's discoverer.

Nonetheless, Walter's travels and strange death kindled Collingwood's ambitions. As Herbert and Bruce took the public-school path that would lead them to Oxford University and beyond, Collingwood plotted a future as an ornithologist. By the age of eleven he could distinguish the songs of most British birds. At fifteen he wrote his first, unpublished book, *English Birds*, complete with illustrations. Westgate-on-Sea was the perfect location for his ambitions. The town is situated on the Isle of Thanet, at one time an island separate from the mainland county of Kent. Over the years the two land masses had joined, when the channel between them had silted up, forming low-lying land that was perfect for spotting birds.

At home, Collingwood was privately tutored, absorbing the foundations of a traditional education: reading, writing and arithmetic. But as he complained later, 'not one of the successive tutors my father employed to educate me showed the slightest interest in any form of natural history'. Even so, Latin proved an easy language for a boy eager to show off his classical knowledge of the derivation of plants and birds, and he learned French with his tutor, Monsieur Le Mullois, on walks along the chalky cliffs and through the watery bogs.

As a boy, Collingwood was particularly influenced by the writings of Henry Seebohm, a Yorkshire-born steelmaker and amateur ornithologist, whose books 'sowed the seeds of restlessness that later made me travel to far lands in search of birds'. One such book told the story of Seebohm's expeditions to the Yenisei tundra in Siberia. Another volume, published in 1890 when Collingwood turned ten, was about Japanese birds – an extraordinary achievement, given that Seebohm had never visited Japan.

Seebohm also wrote in that book about bird specimens obtained by Philipp von Siebold, a German botanist and doctor who had lived from 1823 to 1829 in a tiny Dutch trading community called Dejima, an artificial island within Nagasaki harbour in southern Japan. Given Collingwood's interest in bird migration, he was fascinated by Seebohm's claim that 130 bird species in Japan and Britain were 'absolutely identical or so closely allied that they are not regarded as more than sub-specifically distinct. The birds of Japan do not differ very widely from the birds of the British Islands.'

In Thanet, the months of March and April became particularly important for Collingwood, for it was then that the local birds built nests, laid eggs and fed their hatchlings. Every afternoon the boy would grab sandwiches and a sketchbook and head down to the coast or to Quex Park, a large nearby estate.

Week by week, Collingwood's draughtsmanship skills grew and his drawings evolved into precise illustrations. Inspiration came from one of Britain's most acclaimed artists of the 1890s, Louis Wain, who not only worked at the *Illustrated London News*, but was a close family friend. He occasionally accompanied the Ingrams on holidays and took long walks with Collingwood. In 1895 Wain moved to a house on Collingwood Terrace in Westgate-on-Sea, a street owned by Sir William and named after his son, which was only about 100 yards from The Bungalow. Tragically, Wain later lost his fortune and his mind, moving from one mental asylum to another until his death in July 1939.

As well as watching birds in the Thanet marshes, Collingwood became a regular visitor to Quex Park, an estate that had been owned by the Powell-Cotton family since 1777. In 1894, when Collingwood was thirteen, a young heir, Percy Horace Gordon

Collingwood's first known sketch at the age of eleven or twelve

Powell-Cotton, had inherited the estate on his father's death. Percy was the kind of man whom Collingwood admired greatly. He loved birds, insects, conservation and exploration, and, like Uncle Walter, he had been a Boer War soldier. At Quex, Collingwood tramped the park, unheeding of time, listening to and observing the blackcaps, linnets, nightingales and warblers.

As Collingwood grew into adolescence, a routine emerged in his life. Winter was hunting season, with the local hunt ranging over the flat Thanet fields. He also became a frequent spectator at the popular Thanet coursing meets, where punters betted on greyhounds as they chased hares over a set course. Spring was for birdwatching. August was for grouse-shooting in Scotland and the Lake District with the family. November was for deer-stalking, also in Scotland.

Killing game, Collingwood wrote, was 'a primordial instinct inherent in all men'. At the same time, 'I hope and believe that I have always done so in a sportsmanlike way, which is synonymous to saying that life is taken with an absolute minimum of cruelty.' In September 1905 Ingram nearly lost his own life in an accident while shooting with his father at Barras Moor in Westmorland. The incident became national news – 'Baronet's Son Shot,' screamed the *Manchester Courier* – although it was later reported that Collingwood wouldn't lose his eyesight, as had been feared.

In December, in between all their sporting commitments, the family often decamped to the French Riviera or travelled to Egypt to sail down the River Nile, taking with them some favourite birds. 'Some Egyptian sparrows came on the deck and in some way insulted the English flag,' Mary Ingram recalled about one trip. 'Tony (a sparrow) was infuriated and began a fight. We lost him for five hours. Eventually, our eldest son found him. I only hope that in this mimic bird warfare Tony established the supremacy of England and well punished the Dervish sparrow that had so insulted him.' She added that the crew were given a sheep to celebrate the truant's return, and the boat 'resounded with cries of "Salaam Tony"'.

Collingwood loved sailing. A floating vessel, he wrote, was 'one of the most beautiful man-made objects – one of the few pleasing creations of civilization'. He was also all too aware of its dangers, having heard of his grandfather's death on Lake Michigan and having witnessed two shipwrecks: shortly before his eighteenth birthday, Collingwood was sailing on a yacht off the coast of Margate when a cargo of naphtha on board a nearby iron barque, the *Blengfell*, blew up, killing nine crew members; years later, in November 1910, the 30-year-old Ingram stood on the white cliffs near Dover watching the *Preussen*, the world's only five-masted fully rigged ship, destroying itself on the rocks after colliding with a cross-Channel steamer.

Between family holidays, Collingwood joined his brothers and parents at events in the traditional English social 'season' – horse racing at Epsom, boating at Henley, yachting at Cowes, and cricket at Lord's and the Oval for the elite matches: Oxford vs Cambridge, Gentlemen vs Players, Eton vs Harrow. During frequent visits to London, Collingwood broadened his knowledge of the natural world with trips to London Zoo, the Natural History Museum and the Royal Botanic Gardens, Kew. Few children enjoyed such access and variety at the apex of Britain's imperial power.

By all accounts, Collingwood was a talented, yet precocious adolescent with a strong sense of entitlement that not everyone appreciated. In October 1897, shortly before his seventeenth birthday, he was appointed master of the Thanet Harriers, the local hunt, largely because his father was paying the bills. The hunt committee was less than impressed. They 'rubbed their eyes, adjusted their spectacles and made other actions expressive of their surprise', before telling Collingwood that he was

Collingwood dressed for a hunt, aged about sixteen

too young. Collingwood eventually resigned from the position, after telling the committee that 'if I were not old enough to call the tune, I was certainly not old enough to pay the piper'.

Six thousand miles away, in another verdant tea-drinking island nation, a radically different historical narrative had begun to unfold.

4. Enforced Seclusion

While I was growing up in Tokyo in the 1960s, my grandmother, Katsuyo, who had been born in 1899 and lived with us, would sometimes lament my father's choice of profession. At the time, Papa was a journalist at the *Mainichi* newspaper, a highly prestigious job. Grandma's two elder sons, Masatsugu and

Yukio, had both died shortly after the Second World War, and Hiroyoshi, my father, was her last remaining child. Grandma wasn't satisfied. At least fourteen generations of Abes, stretching back to 1560, had been high-ranking doctors. To her, Papa had broken an almost sacred family tradition by rejecting a physician's life. He had started a medical course in an elite high school in Okayama City in western Japan, but had switched to literature after fainting in the operating room.

Grandma was extremely proud of the family's medical lineage and felt an obligation to her ancestors to continue the tradition. Her late husband, Takatomo, had dreamed of starting a hospital, with his three sons as doctors. He had died of tuberculosis in 1941 aged thirty-nine, having contracted the disease from a patient. 'I can't sleep with my feet towards the ancestors' altar because I'm so ashamed,' Grandma said. 'Naoko, why don't you become a doctor and resume the succession?' As I was a girl, Grandma reminded me, she couldn't force me to; at that time girls were expected to marry, raise a family and not even think about a career. As for Papa … She sucked in her breath and let the moment linger.

After Grandma's death in February 1978, the Abe relatives gathered in Tokyo to pay their respects. Following the cremation, my parents used long wooden chopsticks to remove a few of Grandma's bones from the furnace. 'Ashes to ashes' is, of course, part of the English burial service. In Japan we revere the bones as remains of the deceased. We took them home in a small ceramic urn that was put in a bare wooden box wrapped in a white cotton cloth. In accordance with Japanese funerary rites, the box was placed within the family altar, in a corner of Grandma's bedroom in our eleventh-floor apartment in the

Chōfu district of Tokyo. Mama and Papa later visited the family grave in Yonago, a city on the Sea of Japan, taking the box with them. There a large grey stone monument proclaimed, 'Abe Ke Daidai no Haka' ('Grave of the Abe Generations'). There are no other names.

At the funeral a Buddhist monk lifted the stone, and my parents silently placed Grandma's pot next to those of our other ancestors. They joined the whitened fibulas and powdery femurs of Abes long since forgotten – bones that had run through the Tottori sand dunes; bones that had lain entwined with their loved ones under wild cherry trees in the nearby Chūgoku mountains; bones whose owners had healed the suffering and eased the pain of countless lords and their families; bones that could tell the story of the secretive 200-plus years when Japan was virtually cut off from the world.

What was that history? The short answer is that in the four centuries before 1853, when Japan had a transformative encounter with the West, its history divided in two. The first period, ranging from 1467 to 1600, was the so-called Age of Civil Wars, also known as the *Sengoku* era. The second period, until 1853, was a peaceful time of seclusion called *Sakoku*, which means 'country in chains', when Japan had little contact with the rest of the world. This was the golden age of cherry blossoms.

During the Civil War period a state of anarchy existed throughout much of the country. Different clans, each led by a feudal lord known as a *daimyō*, along with his samurai warriors, fought for power and land throughout Japan's four main islands: Honshu, the largest island; Shikoku, to the south of Honshu; Kyushu, south-west of Honshu; and the diamond-shaped northern island of Hokkaido. Treason, suspicion and assassinations were regular occurrences, even within the same clan. My family belonged to the powerful Amago clan, which reigned over a huge area of western Honshu island. It was challenged by a newly emerging clan, led by a *daimyō* called Motonari Mouri. When Mouri's clan defeated the Amago in the 1560s, the Amago clan split up and my ancestors settled in Yonago, Tottori Prefecture in western Honshu. There the first Abe doctor learned his profession, unaware of the destabilising forces from Portugal and Spain that were threatening Japan and its Asian neighbours.

This was a time when, aided by technological advances in navigation, shipbuilding and gunpowder, the European kingdoms of Portugal and Spain competed aggressively for political, economic and religious influence around the world. To do so, they sent out explorers, resulting in such significant voyages as Vasco da Gama's discovery of a sea route to India in 1498 and Ferdinand Magellan's circumnavigation of the globe in 1521. Meanwhile Christopher Columbus, sailing with the support of Spain's Catholic monarch, had 'discovered' the Americas in 1492, and the Spanish conquistador Hernán Cortés had subdued Mexico in 1519.

The world, most Europeans thought, was up for grabs, to be explored, exploited and colonised at will, with little regard for the local populations. News of Japan's existence had spread. Marco Polo, writing in the thirteenth century after visiting China, had called Japan 'Cipangu, the land of gold'. Although Polo never himself set foot in Japan, his vivid descriptions of its monumental wealth stirred many an adventurer, including Columbus.

Vasco da Gama's voyage to India, the first European journey to Asia by sea, sparked an era of Western imperialism. It was just a question of time before the *Nanbanjin* — as these southern European 'barbarians' were known in Japan — arrived in the land that the Japanese called Nippon or Nihon, the 'land of the rising sun'. The first contact with Japan came almost by accident, when a Chinese junk carrying Portuguese merchants landed on a subtropical island off southern Kyushu called Tanegashima in 1543. It is thought to be the first time that guns arrived in Japan. Other Portuguese merchants followed, establishing a trading centre on Hirado Island, close to Nagasaki in Kyushu. Initially they were welcomed, because Japan's *daimyō* lords wanted muskets, Chinese silk and porcelain, plus other trade. But the arrival of Jesuit missionaries such as Francisco de Xavier in 1549 brought a more powerful threat to the established order: Catholicism. Within thirty years, more than 100,000 Japanese, including many *daimyō*, had converted to that religion, mostly on Kyushu.

The rapid spread of Catholicism, the threat of European colonisation and the continual fighting between rival clans was a toxic mix during the Age of Civil Wars. But near the tail end of the sixteenth century a powerful *daimyō* called Hideyoshi Toyotomi emerged and, through a canny mix of political, social and military skills, unified many of Japan's warring clans. Toyotomi's death in 1598 led to battles between alliances of *daimyō* and samurai from the west and east of Japan in 1600. The east won, led by a charismatic *daimyō* called Ieyasu Tokugawa, who was named shōgun, the top military commander. Shōguns were essentially feudal military chiefs who had ruled Japan since 1192, with different levels of success. Tokugawa's victory gave him control of the entire nation, which he consolidated the same year by setting up his shogunate headquarters in Edo, as Tokyo was originally known.

The *Sengoku* era was over. And for more than two centuries during the *Sakoku* era Japan's islands were calm.

Tokugawa and his shogunate successors handled the problem of growing foreign influence and Christianity with a series of harsh decrees. One, the *Sakoku Edict* of 1635, brought in a series of draconian measures: not only was Catholicism banned, but all missionaries were to be imprisoned, and any Japanese trying to leave the country were executed. When it came to the outside world, trade was restricted to a handful of authorised ports, and all contact with the Portuguese ended. In essence, the country was closed.

By shutting itself off from most of the world and banning Catholicism, Japan avoided being colonised and enjoyed peace for more than 200 years. The Tokugawa shogunate established a nationwide system consisting of about 270 domains, each ruled by a *daimyō*. Although the shōgun led the country, each domain in this feudal system had its own political, economic and social structure. In effect, each functioned as a small country or principality that paid homage to the shogunate. Each domain also maintained a rigid class system. At the top, of course, were the *daimyō*, served by their samurai warriors, who were the only Japanese allowed to carry swords. Beneath them came the farmers and peasants who produced food, followed by artisans who made clothes, swords and other goods. Almost at the bottom were the merchants, segregated and ostracised because they made money from others' labour. Underneath everyone else were the *Eta*: leather-tanners, undertakers and executioners, who dealt with animal slaughter and death.

Throughout the Tokugawa shogunate, the samurai officially remained the upper class, but, as warriors, they had little to do in a time of peace. Over the years Japanese merchants gained wealth and power, and increasingly the samurai and merchant classes intermingled. It was during this peaceful *Sakoku* period that unique arts and culture evolved, mostly in Edo and other large cities. These included *ukiyo-e* woodblock prints, pottery, *haiku* poetry, *kabuki* plays and the creation of about 250 varieties of cherry blossom in the Edo gardens of the *daimyō* lords.

During the *Sakoku* era, the shōguns allowed the Dutch, who had traded in Japan since 1609, to remain, as long as they agreed not to spread their Protestant faith. The condition was that they were confined to Dejima, the reclaimed island off Nagasaki in south-west Japan. It was connected to the mainland by a bridge that the Dutch were forbidden to cross.

The Dutch trading house established on the island acted as a branch of the Dutch East India Company, staffed by fifteen Dutch residents, among whom there was always a doctor. At least three of the doctors stationed on the island over the centuries – Engelbert Kaempfer in the late seventeenth century, Carl Thunberg in the eighteenth century and Philipp von Siebold in the early nineteenth century – were avid botanists, whose Japanese plant collections and descriptions were the first to reach Europe. For Collingwood Ingram, these collections would prove a vital resource.

5. Japan Beckons

Collingwood Ingram's diaries and documents are unclear about what sparked his interest in Japan. Perhaps it was the family's Chin dogs, or reading Henry Seebohm's book on Japanese birds. Or perhaps it was the stories he read in *Tales of Old Japan*, written by a British diplomat, Algernon Freeman-Mitford, who had moved with his family to Westgate-on-Sea in the 1880s. What is certain is that Collingwood's fascination began in the late 1890s and escalated in the early twentieth century.

Over the previous thirty years, since 1868, Japan had undergone a revolution, called the Meiji Restoration. After successfully keeping out the foreign 'barbarians' for two centuries, Japan had flung open its doors to the West, in a bid to avoid invasion and become a modern industrial state. During the closed *Sakoku* period, Europe had undergone a power shift, with Britain, France, Germany and Russia all emerging as strong industrial nations while the Spanish, Portuguese and Dutch fell behind. In the first half of the nineteenth century the Tokugawa shōguns who ruled Japan had felt particularly vulnerable to domination by a foreign power. China, for centuries the most influential power in the region, had fallen prey to Western aggression. After the First Opium War, or Anglo-Chinese War, of 1839–42, the British negotiated the opening of more treaty ports, including Shanghai, where the foreigners were not subject to local laws. China also ceded Hong Kong to Britain. The news of China's semi-colonial status stunned the Japanese, who had admired the so-called Middle Kingdom for more than a thousand years.

But even as the shōguns worried about the British, a new and unexpected threat emerged from across the seas.

This time the menace came from America. In July 1853, four cannon-bearing 'Black Ships' arrived at the entrance to Edo Bay (now known as Tokyo Bay) flying the Stars and Stripes. They were called 'Black Ships' because two steam-powered vessels were belching black smoke. The other two were sailing ships. The fleet's captain, US Navy Commodore Matthew Perry, carried orders from President Millard Fillmore demanding American trade with Japan.

When Perry ordered the ships to anchor off the sleepy town of Uraga and point their cannons at the shore, the Tokugawa shogunate recognised that its future looked bleak. America and the European powers had guns — lots of them. Their technology was advanced compared with Japan's. A samurai sword, no matter how sharp, could not compete with muskets and cannons. The shogunate was forced to sign with

An American 'Black Ship', woodblock print c. 1854

the Western nations a series of treaties that were extremely unfair for the Japanese. Several ports were opened to foreign trade, and Westerners who lived there were not subject to Japanese laws.

Dissatisfaction with the Tokugawa shogunate's inability to resist Westerners' demands prompted a *coup d'état*. Rebel samurai from western and southern Japan backed the benign imperial family in Kyoto as a way to force out the shōgun and open up and modernise the nation on more equitable terms. On 3 January 1868 these rebel samurai deposed the fifteenth and last Tokugawa shōgun, Yoshinobu Tokugawa. The 15-year-old Prince Mutsuhito, later known as Emperor Meiji, became the nation's nominal leader, albeit under the thumb of the rebel samurai, who set up a new government in Kyoto. At the time Mutsuhito had been emperor for less than a year, having ascended the throne in February 1867 after the sudden death of his father.

All these tensions led to a civil war, called the Boshin War, which began in late January 1868, less than three years after the American Civil War had concluded. In July the city of Edo fell to forces loyal to the emperor. The new government promptly changed its name from Edo to Tokyo, as the de facto capital, and proclaimed a new era, called *Meiji*, which meant enlightened governance, even before the civil war ended. The emperor then moved from Kyoto into a 400-year-old castle in Tokyo where the successive Tokugawa shōguns had lived. The castle is now known as the Imperial Palace. The Boshin War conflicts in 1868 and 1869 between the imperial forces and the deposed shogunate proved an unequal contest, for the imperial forces were not only backed by the British, but also had more men and arms.

The leaders of this new government knew that, in order to survive, they needed to establish a modern and united nation-state as quickly as possible. In quick succession the government forced the *daimyō* lords to give away their property to the emperor. Bureaucrats from Tokyo now took the place of the *daimyō*. The government also abolished the class system, leaving thousands of samurai without jobs. These samurai became mercenary *rōnin* – samurai without masters – after the downfall of their feudal lords. From Kyushu in the south to Hokkaido in the north, the nation's new rallying slogans became 'Rich Country, Strong Military' and 'Leave Asia, Join Europe'. Catching up with the West became a national obsession and a new era of rapid economic, social and political development took hold.

After centuries of discouraging contact with most foreigners, Japan welcomed thousands of Western educators, entrepreneurs, government officials, naturalists and adventurers. Between 1868 and 1900 the Japanese government hired about 8,400 foreigners to advise businesses, the military, lawyers and bureaucrats.

Half of these foreigners were British; the rest were mainly German, French and American. An Englishman named Josiah Conder, for example, was hired in 1877 to teach architecture at Tokyo Imperial University, where he helped to transform the capital's Marunouchi area into a London-style business district. Conder also fell in love with Japanese gardens and with *ikebana*, the art of Japanese flower arrangement, writing influential books on flowers and floral arrangement. Obsessed with all things Japanese, these Westerners returned home with unknown plant and animal specimens, plus a thousand and one tall stories.

The Japanese people 'looked as if they had walked out of a fairy tale', wrote Marianne North, a British botanical artist who arrived in November 1875. Her sentiment was echoed by Lafcadio Hearn, a nomadic Greek-born journalist who arrived in Yokohama in April 1890: 'The stranger finds himself thinking of fairy-land – a smaller and daintier scale, a world where all movement is slow and soft, and voices are hushed – a world where land, life and sky are unlike all that one has known elsewhere.'

Such was the explosion of interest that, on both sides of the Atlantic, Japanese arts, crafts and culture became a craze after the 1860s. In particular woodblock prints and paintings featuring cherry blossoms, by artists such as Katsushika Hokusai and Utagawa Hiroshige, influenced and motivated painters like Edouard Manet, Claude Monet and Vincent van Gogh.

The centre of this so-called 'Japonisme' craze was Paris, where Japanese art, design and fashion were incorporated into Western aesthetics, but the fascination quickly spread to London, New York and beyond. In 1875 the prestigious Liberty department store opened on Regent Street in London's West End, specialising in ornaments, fabrics and art objects from Japan, India and the Far East. The following year Christopher Dresser, a designer, visited Japan for four months as a representative of the South Kensington Museums. He also bought Japanese goods for Tiffany & Co.'s store in Union Square West in Manhattan.

It was against this background that Collingwood Ingram was slowly drawn into Japan's embrace. On Wednesday, 24 June 1896, the 15-year-old attended a special matinee of Gilbert and Sullivan's comic opera *The Mikado* at the Savoy Theatre.

Collingwood loved the performance. W.S. Gilbert, the lyricist, had set many operas overseas, as a way to poke fun at British politics and its institutions without causing too much offence. He set *The Mikado* in the fictitious Japanese town of Titipu.

For authenticity, Gilbert visited a Japanese exhibition in Knightsbridge and recruited Japanese workers, who instructed the actors in mannerisms and culture. So when Nanki-Poo, Ko-Ko and other cast members sang 'The flowers that bloom in the spring, tra la', they did so under a set design of weeping cherry blossoms. Less than a month later Collingwood and his mother went to Daly's Theatre in Leicester Square to see *The Geisha*, a musical comedy set in a Japanese tea house. It featured songs such as 'The Dear Little Jappy-Jap-Jappy', 'Chin Chin Chinaman', 'A Geisha's Life' and 'The Interfering Parrot'. Collingwood's diaries don't record what he thought of this Asian mishmash, but he appears to have been entranced by the call of the Orient.

That same year, 1896, Collingwood's estranged grandmother, Ann, died at her riverside home at Walton-on-Thames, west of London. She was buried in Boston, Lincolnshire, next to her first husband, Herbert. Collingwood's father and uncle inherited nothing. Instead, Ann's huge estate was divided equally between her five daughters, and lucrative trusts were drawn up for their children and grandchildren.

Collingwood himself rarely mentioned his wealth in his diaries and books, but there's no doubt that he was financially secure for life. With the means to travel wherever he wanted, the only question was where? The Grand Tour of Europe that so many well-to-do young British men traditionally embarked upon held little appeal for him. After all, he already knew more of the world than most people of his age, having visited Egypt several times, lived on the French Riviera, taken yachting expeditions to Corsica (accompanied again by Tiny, the albino sparrow) and spent time in European capitals.

It was only natural to consider more exotic locales. Not only had his mother's accounts of growing up in the wilds of South Australia made a strong impression, but her eldest brother, Edward Charles Stirling, was a well-known anthropologist, artist and explorer there. These were all interests that Collingwood shared. When he officially became an adult and gained the right to vote in October 1901, it was the perfect time to visit the colonies and the outer world.

A self-portrait by Ingram, 1899

It was a new beginning for Collingwood's brothers, too. After graduating from Oxford University, Bruce replaced his father, Sir William, as editor of the *Illustrated London News* in 1900, a position that he retained until his death in 1963. Meanwhile Bertie became a man of leisure, enjoying sports, foreign trips and collecting Chinese and Japanese artefacts.

It was also a new era for Great Britain and the increasingly prosperous United States. Queen Victoria's death in January 1901, after almost sixty-four years on the throne, marked a historic juncture between epochs. Victoria had been mourning Prince Albert for almost forty years. When her eldest son, King Edward VII, moved into Buckingham Palace with Queen Alexandra and their own collection of Japanese Chin dogs, the country's mood lightened. The United States, meanwhile, welcomed a new head of state. On 14 September 1901 Theodore Roosevelt was sworn in as the US's twenty-sixth president, following the assassination of William McKinley.

6. The Rising Sun

I have never seen man and nature in such close accord or a land of such artistic taste. Nature is not always an artist; at times she is too lavish with her colours, at others she is arrogant or she may be overproud. Not that nature errs frequently when left alone. It is usually man that defaces her work, that scars her smiling features with smoke-begrimed cities.

Untrue as it may sound to the uninitiated, here (in
Japan) man adds to, instead of detracts from, the beauty
of his country.

Collingwood Ingram's first glimpse of the nation that
would for ever influence his life was of the south-western
port of Nagasaki in the thirty-fifth year of the reign of Emperor
Meiji. To Ingram, it was simply Friday, 5 September 1902. A
mature, confident and well-travelled 21-year-old, he disem-
barked from Nippon Yusen Line's *Kumano Maru*, a Japanese
passenger ship, with a friend from Westgate-on-Sea, Harold
Cobb. The two had embarked on the three-week voyage from
Townsville, a city in north-east Queensland, after travelling
1,200 miles from Stirling, a township in South Australia close
to Adelaide, which had been named after Collingwood's late
Scottish grandfather.

Ingram and Harold had stayed for two months in Australia,
most of the time with Collingwood's uncle, Edward Stirling —
the first Professor of Physiology at the University of Adelaide
— at St Vigeans, a 6½-acre property named after the Scottish
village where Edward had attended school. Stirling was a typical
Victorian Renaissance man, steeped in both the arts and the
sciences. He was particularly well known in South Australia for
a 2,000-mile north–south trek that he had made in 1890 from
Port Darwin to Adelaide to collect fauna and flora.

The timing of Ingram's arrival couldn't have been better.
Japan had stunned the world in 1894–5 by defeating China in
the Sino-Japanese War. The victory proved to the Meiji government that its bold bid to 'Catch Up with the West' was bearing
fruit. The United Kingdom had approached Japan as an ally,

looking to preserve its influence in Asia, and in January 1902 the two countries had signed the Anglo-Japanese Alliance. This ended Britain's so-called 'splendid isolation' policy since 1866 of avoiding foreign alliances. Designed in large part to oppose Russian expansion in the region, the pact was extended in 1905 and 1911.

Japan, and not China, was now the region's dominant force. 'The war opened the eyes of the foreigner to the fact that as far as strategy was concerned, the Japanese had really profited by their studies ... at the hands of their European and American instructors,' wrote Stafford Ransome, a special correspondent of the *Morning Post* newspaper, in 1899.

Even the ship on which Ingram had travelled to Japan underscored the nation's lofty expectations. The *Kumano Maru*, built for Nippon Yusen in Glasgow's Fairfield shipyards, had been launched in autumn 1901 to compete with Western passenger vessels plying their trade between Asian ports. Townsville–Nagasaki was a new route, reflecting Nippon Yusen's global ambitions. Ingram and Harold were among the first Western passengers to make this journey.

In Japan, Ingram's itinerary followed the route that small groups of adventurers had established since Japan had started allowing in foreign visitors four decades earlier: Kobe, Osaka, Kyoto, Hakone, Tokyo, Kamakura, Nikko and Yokohama. All were popular tourist destinations on Honshu, the most populous and largest of Japan's four main islands. Far fewer visitors went to Kyushu, Shikoku or snowy Hokkaido, not least because the distances were so great. Still fewer went to the tropical islands of Okinawa, south of Kyushu.

On a globe, at a glance, Japan doesn't appear much larger than Great Britain. In fact it stretches in a crescent shape for 1,800 miles from north-east to south-west — three times Britain's length, and about the same distance as from London to Moscow or from Maine to Florida. Transport links in Japan were limited in 1902, in part because more than three-quarters of the country was mountainous and most of the population lived in major cities on a fertile coastal plain in the south of Honshu. But Ingram cared little for the cities. He couldn't wait to leave the population hubs for the countryside, where he was enchanted by the scenery, the people and their customs:

> The country seems steaming damp and in consequence
> deliciously green and mossy. From the cherry trees there
> comes a continual flow of sound from some loud-noise
> insect [the cicada] — it is like an auctioneering frog,
> singing out in a loud humdrum voice 'Going! going!
> gone!' Not that I am acquainted with an auctioneering
> frog, but imagine one conducting a sale and there you
> have our friend of the cherry trees.

Ingram found the Japanese landscape 'becoming — tidy cheques of paddy fields, bamboo groves, little villages and usually a clear, swift-running stream'. In the cool hills above Hakone, a hot-springs resort near the towering Mount Fuji, 'plants grew in ravishing luxuriance; cherry, maple, pine and elm in one green confusion of foliage, while every short while there came a gushing waterfall down the hillside, bringing with it a draught of cold, vaultish air with the scent of moss and earth'.

With fellow foreigners from the ship, Ingram was carried in wooden rickshaws up and down hills and visited enough shrines and temples, he believed, to last him a lifetime. A lengthy *kabuki* play in Kyoto was interesting mostly for the sight of the audience waving small, half-moon-shaped paper fans. 'To see the hundreds of fans rising and falling was like the ripple of wind upon a corn field,' he wrote. He also found Japanese women 'undeniably fascinating, one continual ripple of mirth'.

One day, while his fellow travellers remained in Tokyo, Ingram went alone to the tiny island of Enoshima, 50 miles south of the capital, and walked for hours along the beach with an elderly smiling monk who 'explained everything in Japanese and was highly pleased when I nodded fatuously by way of acknowledgement'.

After just fifteen days in the country, Ingram was smitten. On 20 September 1902, when he boarded the SS *Peru* at Yokohama for San Francisco, he became sentimental: 'My visit to Japan has been so enchanting that I have had no time to do else than stand agape and watch the different vistas pass away without record in my journal. But a brief fortnight has left me with more memory pictures than months of travel elsewhere.'

As the vessel steamed away from the port, Ingram looked back and saw the snow-tipped Mount Fuji rising in the distance: 'When the land was behind the horizon this pyramid still shadowed against the evening sky – a lone mark upon the ocean, almost alone in the heavens, alone save for a few gilt-edged clouds that sailed in the sea of orange.'

From the moment Mount Fuji disappeared quietly into the darkening sea, he knew that his infatuation with Japan had only just begun.

7. The Birds and the Bees

Five years later, Collingwood Ingram was back in Japan. Between his first and second visits Japan had fought – and won – another confidence-boosting war. This Russo-Japanese War of 1904–5 was primarily fought over the two nations' rival ambitions in Korea and Manchuria.

Britain had had a strong influence on that war, owing to the 1902 Anglo-Japanese pact. Most of Japan's ships, for instance, had been built in British shipyards, and the naval officers who commanded them were trained in Britain. Loans from British banks had enabled Japan to pay for supplies. The war itself ended with the Treaty of Portsmouth, mediated at the Portsmouth Naval Shipyard in the state of Maine by the American President Theodore Roosevelt, who was awarded the Nobel Peace Prize for his efforts. Like the hostilities with China ten years earlier, the war further transformed the regional balance of power and added to Japan's global esteem.

There were also changes in Ingram's life: when he arrived in Japan for his second visit on 20 April 1907, he brought with him his bride, Florence.

Determined to return to Japan, Ingram had convinced Flo (as she was always known) that they should honeymoon there in the spring, after their October 1906 wedding. For her, the trip proved a honeymoon from hell. She was more than two months pregnant and fell sick after being cooped up in a ship's cabin for weeks. The pair had met after Ingram's first Japan trip, and they appeared well suited. Florence was from a prosperous family with deep roots in Scotland. Her father, Henry Rudolph

Florence Ingram

Laing, founded a firm of stockbrokers called Laing & Cruickshank. Her grandfather, Samuel Laing, was a Liberal Member of Parliament, like Ingram's father and grandfather. Ingram and Florence married at Holy Trinity Church in Chelsea a few days before his twenty-sixth birthday. After the wedding they had moved to a detached house in Westgate-on-Sea, close to The Bungalow. Well-heeled and well-connected, the young couple could indulge their interests, of which ornithology and oology – the study and collection of birds' eggs – were top of Ingram's list.

On this Japanese trip Ingram's ambition was to find birds' eggs and capture birds whose skins could be returned to England.

In particular he wanted to become the first Englishman to find the eggs of a species called White's Thrush, which was rarely seen in Britain. The bird was named after the Reverend Gilbert White, the eighteenth-century English naturalist whom Ingram revered. Like many fervid collectors (whether of stamps, coins or art), Ingram became almost obsessive about tracking down these elusive birds.

After leaving the ship, Collingwood and Florence travelled to Tokyo to apply for a government permit to capture birds. The paperwork took three weeks to obtain, even though Ingram was aided by Isao Iijima, Professor of Zoology at the Imperial University of Tokyo, an English-speaking bird expert who was so well connected that he frequently went pheasant-hunting with the emperor.

Permission obtained, Ingram left his bride in Tokyo and departed on a three-week walking and collecting trip in the foothills of Mount Fuji. With help from locals – whom Ingram paid according to the rarity of the species they discovered – he eventually found a well-camouflaged, moss-covered nest of a White's Thrush on a half-fallen tree.

Florence had little interest in her husband's passion. Now three months pregnant, she rested in Tokyo hotels while her new husband based himself in a tiny village called Subashiri and wandered the eastern slopes of Mount Fuji. 'Grandpa did whatever he wanted. She just adjusted to his needs,' Veryan Pollard, Collingwood's granddaughter, told me. 'Grandma was a long-suffering woman, a very traditional lady who didn't complain and accepted her lot in life.'

After almost two months in Japan and the victorious discovery of some White's Thrush eggs, the couple began their journey home from Japan by ferry across the Sea of Japan to Vladivostok. There, anxious to safeguard Florence and the unborn child, they boarded the newly completed Trans-Siberian Railway rather than attempt a lengthy voyage, and crossed the steppes of northern Manchuria to Moscow and Berlin. To Ingram, the honeymoon had been an overwhelming success. He had found seventy-four kinds of Japanese bird, including a White's Thrush, and returned home with drawings, eggs and other birding mementoes. These formed the basis for a paper that he wrote for the January 1908 edition of *IBIS*, the prestigious magazine of the British Ornithological Union.

Ingram and Florence arrived home in Westgate-on-Sea in the summer of 1907 after their three-month sojourn. There, on 1 November 1907, Florence gave birth to their first son, Ivor Laing Ingram. Seventeen months later a second son, Mervyn Jeffry Ingram, was born, on 21 March 1909. A third, William Alastair Ingram, followed on 26 August 1913. Finally, on 4 January 1917, their only daughter, Certhia, was born. She was named after a Eurasian bird, a tree-creeper with the Latin name of *Certhia familiaris*.

As Florence and the family's nurses and maids raised the children, Ingram concentrated on building a name as an ornithologist. At the Natural History Museum in London, for instance, he helped to curate bird skins sent from Australia by a naturalist, Wilfred Stalker, on commission from Sir William. Then in January 1913 he travelled to a remote 400-acre island in the West Indies where his father, now in retirement, had begun an ambitious project to prevent the extinction of a species.

Sir William had heard that the Greater Bird of Paradise, the largest member of the colourful breed found in the rainforests of New Guinea, was endangered because its iridescent, maroon-streaked tail feathers were in such high demand as plumes for ladies' wide-brimmed hats.

To prevent further depletion of the birds' numbers, in 1908 Sir William bought Little Tobago, an uninhabited Caribbean island off Tobago, and hired a naturalist to collect live specimens of this rare bird. In 1909 forty-seven young birds were shipped from New Guinea to the mile-long island aboard a German liner and released. To look after the birds, Sir William hired a middle-aged Swiss sailor who went by the name of Roberts. Every few weeks, Roberts sent Sir William an update about the birds, along with watercolour sketches and crude poems, which he apparently wrote 'in a state of helpless inebriation'.

Always ready for an adventure, Ingram lived for a few days with Roberts in a wooden shanty hut on a bluff near the sandy landing beach. Early one morning they heard the Greater Bird of Paradise's distinctive *wauk-wauk-wauk* cry and soon found four of the birds eating pawpaw fruit. Ingram wrote a research paper in 1913 about this wildlife colony, which survived on the renamed Bird of Paradise Island until the 1960s. It is unclear why the colony died out, but the Bird of Paradise still features on Trinidad and Tobago's $100 note. Poor Roberts died from pneumonia shortly after Ingram's visit, after his boat capsized in the surf and he collapsed in a drunken slumber in his hut.

Ingram was captivated both by Little Tobago's beauty and by his father's fascinating experiment with the birds. But any plans to return to the Caribbean were put on hold by the assassination on 28 June 1914 of Archduke Franz Ferdinand of Austria, the heir to the Austro-Hungarian throne — a murder that propelled England into what became the First World War. Few people in England expected the conflict to last beyond the end of the year. The Thames estuary near Westgate-on-Sea became an important strategic location, because it housed the Chatham and Sheerness naval bases.

In 1914 the Royal Naval Air Service opened a seaplane base near the town, and later an inland airfield, today called Manston Airport. Ingram, then thirty-four, was commissioned into the Kent Cyclist Battalion, whose members watched the southern coast for possible enemy incursions and reported sightings of planes and Zeppelins. For Ingram, the commission was perfect, as their infantry battalion spent much of their time in Romney Marsh, where he could watch an abundance of migrant and breeding birds while still seeing Florence and his young children. These good times were not to last.

8. Ingram's War

Collingwood Ingram's war began in earnest early on Thursday, 7 December 1916, when he arrived in the northern French port of Boulogne on an officers' boat flanked by two destroyers. In London, on that same chilly day, David Lloyd George became prime minister of a coalition government.

Earlier that year Ingram had left the Kent Cyclist Battalion, after being seconded to the Admiralty to learn how to adjust magnetic compasses in aircraft. For pilots who flew over enemy lines, a precise compass was a necessity, particularly at night, in fog or in cloud cover.

After training, Ingram became a captain in the Royal Flying Corps and was ordered to report to an aircraft-repair depot at Saint-Omer in the north of France, about 30 miles from the nearest trenches. This depot serviced the British aerodromes west of Ypres, the strategic Belgian town where more than 850,000 Allied and German troops lost their lives in brutal battles between 1914 and 1917. Ingram's war journal from those days is an extraordinary document.

'The captain', as he was fond of being called, filled five leather-bound books and six sketchbooks with meticulous pencilled notes and detailed drawings about his life near the battlefields of northern France. Aside from adjusting compasses, Ingram indulged in a frenzy of activity, as though every day might be his last. He rode horses through the French countryside. He accompanied daredevil pilots on hair-raising joyrides in Sopwith Camel planes. He hunted wild boar and shot partridges, rode in hot-air balloons, visited museums and bordellos, and ended the war in the German town of Cologne.

His diaries rarely mentioned his wife and children, although a month after arriving in France, Ingram flew home for three days to see Florence and their newborn daughter, Certhia. As well as raising her children during the war, Florence worked in Westgate-on-Sea as a VAD (Voluntary Aid Detachment) nurse alongside her sister-in-law, Hilda Ingram, and military nurses treating soldiers who had been wounded in France and Belgium.

Back in France, Ingram made field notes about the 170 different varieties of bird that he observed in the area: the abundance of jackdaws, sparrows, blackbirds and rooks in the tall trees near a chateau that he shared with fellow officers; tree-creepers in the forest of Chaudeney; crested larks on an aerodrome; and bullfinches in birch trees. In many of these entries it seems as if the war didn't exist.

Compared with most British troops, Ingram spent the winters of 1916 and 1917 in conditions of relative luxury, away from the freezing trenches in which millions of Allied and Central Power troops lived and died. He found great beauty in the frozen French countryside, writing on 30 December 1917:

> A moon of unsurpassable brilliance flooded the silent landscape with a cruel glare of greenish light, which traced sharp inky shadows of the trees on the rounded white folds. The snow crystals caught and reflected the moonlight upon a myriad facets until I appeared to be walking in a world of sparkling diamonds.
>
> The frightful stillness of the woodland at midnight was almost startling – everything seemed to be frost-bound and nerveless. Even the icy air seemed frozen into immobility. The crisp crunch of my footfall appeared to be an unpardonable intrusion, while the scars they made upon the smooth field of scintillating white seemed a positive sacrilege.

Yet his journals also told another story, and gave insights into choices that he made after the war. For Ingram was horrified by 'the uncared-for dead, lying upon fields, mutilated and disfigured beyond recognition by many months of shell fire':

> The year-old rat-eaten corpses, the shrivelled dismembered limbs, still booted or clothed, the half-bare

skulls — these are not subjects to dwell upon — but they tell their tale of heroism — for surely only heroes or madmen would have attempted to cross those open stretches of pitted ground between those ugly barriers of barbed wire.

It is a commonplace platitude to say that modern warfare is devoid of romance. Romance indeed? It seems to me that war is merely legitimisation of murder and (in the case of the enemy at any rate) the sanctioning of all the more brutal instincts of mankind.

Along with the endless human suffering, Ingram was outraged by the destruction of nature: 'The most galling evidence of wanton devastation was the fact that, without exception, the trunks of all the fruit trees had been cut through at their base. The beautiful poplars and elms that formed the roadside avenues had suffered in a like manner. Even the rose bushes in the chateaux' gardens had been ruthlessly hacked to pieces. In short, anything that was either beautiful or useful had been brutally manhandled.'

The ravages of war have laid the country waste and now the rolling hills are nothing more than a treeless and lifeless expanse of rank weeds, shell-holes, trenches and graves — with here and there a derelict tank to break the dreary monotony.

This war-stricken belt, with its shattered leafless trees, suggested to me a country overrun by some vast horde of devastating insects — a plague of monstrous locusts that had devoured all living things and left behind nothing but an ugly pock-marked waste.

By the late summer of 1918 the Central Powers were on the run. After the failure of Germany's so-called Spring Offensive and the Second Battle of the Somme, victory loomed for the Allies. By November, Ingram's war was over.

9. Birth of a Dream

On Armistice Day, 11 November 1918, Ingram had just turned thirty-eight years old. He was alive, unlike ten million military personnel and seven million civilians from all sides of the conflict, including dozens of his friends. His limbs were intact. He was exceptionally lucky not to suffer from shell shock or any other apparent psychological affliction.

Yet he was troubled. For the past two years Florence had quietly brought up their four children, now aged one, five, nine and eleven, in Westgate-on-Sea without him. His brothers, Bertie and Bruce, were now forty-four and forty-two respectively. They were settled, with Bertie now a collector of Asian artefacts and Bruce the editor of the *Illustrated London News*. But what should he do? Ingram had already travelled the world, become an eminent ornithologist and contributed to the war effort. He was independently well-off, with supportive parents and brothers. Yet it wasn't enough. He seemed to be experiencing what today might be called a mid-life crisis. Like millions of other British soldiers, Ingram sought to adjust to civilian life and to the social and economic changes that were rocking the nation.

Fewer than twenty years since the end of its Victorian heyday, Britain was struggling economically, having borrowed billions of dollars from the United States to fund the war. Its empire was starting its slow decline. As the international landscape changed, so the US was emerging as a new superpower. And Japan, too. During the war Japan had allied itself with the Entente Powers against Germany, expanding its influence in China and taking over Germany's territories in the Mariana, Caroline and Marshall Islands in the Pacific Ocean. By 1918 Japan was supplying goods to its European allies, and a surge in the nation's exports had led to an industrial boom. Its reward was a seat at the 1919 Versailles Peace Conference, followed by a permanent seat on the Council of the League of Nations.

Ingram himself was at a loose end. Despite his love of birds, he was disillusioned with his vocation. There were too many ornithologists, and nothing new to discover. 'Ornithology has become – let us admit it – a somewhat tired and exhausted science,' he wrote, criticising its focus on details such as the size, geographical distribution and taxonomy of birds. 'The real bird lover only looks at these things as unavoidable evils. He is chiefly concerned with the living creature. It is they [the bird lovers] who lead you forth into the sunny fields and shady woods to hear sweet music.'

One research paper that he read at the time pushed him over the edge. 'When the editor of one of the world's premier ornithological journals deemed it of sufficient interest to publish a paper in which the author recorded the number of times a great tit defecated every 24 hours, I came to the conclusion that it was high time I occupied my thoughts with some other aspect of nature,' he wrote. 'I chose plants.'

Ingram's decision coincided with the family's move away from Westgate-on-Sea. In early 1919 they found a new home to accommodate their fast-growing family, maids, housekeepers and nanny, in Benenden, Kent, about 50 miles south-east of London. A once-sleepy farming village, Benenden had become fashionable as a second-home destination for well-to-do businessmen and politicians. A railway station had opened in the 1840s at Staplehurst, just seven miles away, which connected the Kent villages of the area to both London and the south coast.

Much of Benenden and its surroundings had been purchased in 1857 by Gathorne Hardy, a Yorkshire-born owner of an iron works in Bradford in the north of England. Hardy became a Conservative Member of Parliament in 1856 and a peer in 1878, taking the name Lord Cranbrook. Much as Sir William had done at Westgate-on-Sea, Hardy poured money into the village of Benenden, restoring its church, building schools and overhauling Hemsted House, his family's main residence. His tenants included many local dairy, sheep and arable farmers. In 1893, Lord Cranbrook commissioned a half-timbered house called 'The Grange' for his unmarried daughter. After his death in 1906, the estate was sold to Harold Harmsworth, founder of the *Daily Mail* and *Daily Mirror* newspapers, who became Baron Rothermere of Hemsted in 1914. But after the baron's two eldest sons, Harold and Vere, were killed in the war, he disposed of his Benenden property. The Grange, along with two large farms, was sold to the Ingrams.

The Grange was a quintessentially upper-middle-class English home, situated next to a church, St George's, and Benenden's village green. It was built in Tudor-Gothic-style architecture, with black beams on white walls and tall brick chimneys. Through the entrance hall were a dining room, a sitting room, a library and a morning room, all 12 feet high and heated by hot-water radiators. Up an oak staircase were seven bedrooms and dressing rooms for the family, as well as a day-nursery. On the floor above there were ten servants' bedrooms and boxrooms. Above that was the attic, which became Ingram's hideaway.

Outside the house stood a boot room, coal and wood stores, stables and an empty gardener's cottage. But no real garden. When the Ingrams moved to the property in the summer of 1919, they found only an abandoned rosebed and a few neglected shrubberies surrounding what had once been a tennis court. Beyond a grass field and behind a belt of trees there was a derelict kitchen garden. The rest of the property was farmland and meadows, where sheep and goats grazed.

For Collingwood Ingram, this was ideal. He decided to create a different kind of English country garden out of one of the meadows south-west of The Grange, which would include both native and exotic trees and shrubs. His goal was to produce 'a succession of sylvan glades, leaving a series of open grassy spaces, each one of a different size and shape'. 'Each was designed to terminate at its furthest end in a sharp bend, the purpose being to close every vista in order to intrigue the eye and to make a stranger wonder what new treasures awaited him round the hidden corner.'

As he had hoped, this new preoccupation began to wean him 'from what had been up to then a lifelong pursuit – ornithology'. Plus, the dilapidated garden that Ingram had taken on hid a secret: two ornamental cherry trees.

One lay to the west of the house, the other to the rear. They were mature trees, about twenty years old, which had probably been planted in the mid- to late 1890s by Lord Cranbrook's gardeners, when cherry trees from Asia were a rarity. The English were, of course, familiar with cherries that bore fruit, but much less so with flowering cherries. The cherry trees weren't blossoming when the Ingrams moved in, but the long and sturdy branches had deep-green leaves that attracted Ingram's attention and reminded him of his visits to Japan.

In the spring of 1920 these two trees became smothered in silky pink blossoms and deposited a carpet of wispy, weightless petals beneath them. Viewing the blossoms on the largest tree, which was 25 feet high with branches that spread 42 feet, Ingram wrote that it 'would be difficult to conceive a more striking floral display'. The trees' strong trunks, branches, leaves and, above all, blossoms were a testimony to nature's bounty.

The Hokusai *in bloom at The Grange in the early 1920s*

It wasn't long before Ingram became preoccupied with this new interest. Japanese cherries were virgin territory in Britain, both for research and for collection. It was a perfect match for a man seeking a new life for himself in the aftermath of so much death and destruction. 'This was the beginning of the love story between Collingwood and the cherry blossom,' Ernie Pollard, Ingram's grandson-in-law, told me.

Within months Ingram's goals began to crystallise: to collect as many cherry-tree varieties for his garden as he could find, and to become a globally recognised cherry-blossom expert.

PART TWO

—

CREATION AND COLLECTION

10. Twin Quests

In the early 1920s, seated in the paper-strewn attic of The Grange, his home in Benenden, Collingwood Ingram pondered the enormity of the task he had set himself. It was one thing to create a beautiful garden full of cherry trees, quite another to become a cherry connoisseur with a diverse collection. Ingram knew little about cherries, or indeed about gardening. In Westgate-on-Sea his father had employed a team of gardeners to create and maintain the grounds at his numerous properties, while Collingwood wandered the woods and marshes looking at birds.

As an ornithologist, he had clashed with the experts whose views he opposed or derided. With cherry trees, there were few specialists in the West to contradict his findings. So much the better. And so in 1920, setting aside his extensive ornithological research, Ingram started reading anything and everything about his new passion, and about Japan. His first two visits, in 1902 and 1907, intrigued him, in part because of what he called 'the marked kinship between the fauna and flora of the two countries', by which he meant Britain and Japan.

Because there is a certain similarity in their insular climates, many parallel forms of plant and animal life have evolved. This resemblance has a certain fascination.

One meets, as it were, old friends in a new guise. They are the foreign counterparts of things one has learnt to know and love at home – the same, and yet not the same. Possibly it is for this reason I have always been attracted to Japan.

Indeed, cherry blossoms themselves were 'the same, and yet not the same' in the two countries. For several thousand years about 100 different wild species of cherry tree had grown around the world, mainly in the temperate regions of the northern hemisphere, from the foothills of the Himalayas to the northern Italian Alps. Many species grew in Russia and China, and a handful in Europe and North America.

Until the late nineteenth century Japanese cherries were not particularly fashionable imports in the West. At that time there was more demand for azaleas, chrysanthemums and rhododendrons, in part because Japan had been closed to most outsiders, but largely because flowering cherries did not produce edible fruit and many Europeans had a fixed notion that a cherry tree should produce something to eat, like most local fruit trees. Hence the reason the English called Japanese cherries 'pseudo-cherries' and the Germans called them 'false cherries'. The Latin name was *pseudo-cerasus*. These were hardly attractive descriptions for Western horticulturalists. Thankfully, that negative image gradually disappeared in the late nineteenth and early twentieth centuries, as journalists and other visitors to Japan published first-person accounts of the ornamental cherries' beauty.

Interest in Japanese cherries also benefited from the growth of gardening as a leisure pursuit in the late-Victorian and

Edwardian eras. This was a time when gardening clubs and horticultural societies proliferated, with the nouveau riche especially eager to obtain, grow and display rare Oriental plants that didn't exist in northern Europe because it was too cold. The problem was that many exotic plants needed to be kept warm during the British winter, which only became possible after new glass-making techniques were perfected in the 1830s. That, together with the abolition in 1845 of the tax on glass, enabled cheap production of glasshouses.

Like that of many of his fellow gardeners in Britain, Ingram's gardening philosophy was influenced by William Robinson, a self-educated Irishman whose two best-selling books *The Wild Garden* (1870) and *The English Flower Garden* (1883) preached the gospel of nature's glory and supremacy. The gardener, Robinson believed, was simply someone who should assist nature to express herself, which meant that the garden should eschew pretension, excluding 'showy labels' and even Latin flower names. 'A garden should be a living thing,' he wrote. 'The most beautiful effects must be obtained by combining different forms so as to aid each other and give us a succession of pictures.' Ingram, whose goal at Benenden was to create 'a succession of sylvan glades', couldn't have agreed more.

The 'pictures' that Ingram wanted to produce at The Grange all involved interspersing Japanese cherry trees with other plants and shrubs. As he sketched out plans for his dream garden, he sought more information about the history and culture of the cherry trees that would be the focal points of his property. What he discovered, in brief, was that cherry trees had been part of Japanese people's lives for more than 2,000 years. As cherries were among the first trees to blossom in the Japanese spring, they had come to serve as bellwethers to farmers. When the cherries bloomed, the farmers believed, it was time to start planting rice seeds, Japan's staple diet.

As a cherry-blossom nation, Ingram learned, Japan's abundance of cultivated, or man-made, cherries was unique. No other peoples in the world cultivated cherries to such an extent. All of these man-made varieties were derived from only ten known species of natural wild cherry that grew in Japan. Each of these species grew in a different location, although there was a lot of overlap. The most popular wild cherry was the Japanese mountain cherry, known as *Yama-zakura*. Widely seen in the mountains of central, western and southern Japan, where the climate is mild, it bore five-petalled pinkish-white single blossoms and was the most celebrated type of tree in Japan. It was the species that Japanese authors, poets and playwrights usually referred to until the late nineteenth century, when another cherry, the cultivated *Somei-yoshino* variety, became the dominant *sakura*.

Another wild species was Sargent cherry (*Ōyama-zakura*), which produced crimson blossoms and grew in abundance on the northern island of Hokkaido and in northern Honshu, Japan's main island. One of Ingram's favourites, it had been named after the American plantsman Charles Sprague Sargent. Meanwhile the umbrella-shaped Fuji (*Mame-zakura*) species grew around Mount Fuji in central Honshu, producing small white to light-pink flowers. Another species, Ōshima, produced large white flowers and was native to the wind-swept and rainy Izu islands that stretched out into the Pacific Ocean south of Tokyo.

Japan's well-defined seasons, along with plentiful rain and soil that was rich in volcanic ash, combined to help the cherries flourish. Additionally, in Japan's predominantly mountainous landscape, steep slopes led to sharp differences in temperature between areas close to each other, offering an ideal environment for multiple species of cherries to grow and naturally hybridise with each other. For example, in the Abukuma mountains of central Honshu, two different species — *Yama-zakura* and *Edo-higan* — grew in the vicinity, creating different forms of cherries as a result of their union. Whenever someone found a beautiful cherry in the mountains, they would collect its seeds and plant them in their village, take a cutting from the tree and graft it or simply uproot the entire sapling and replant it where they lived.

From these ten wild species, more than 400 flowering varieties had been cultivated by humans in Japan over the previous 1,200 years. Domestic cultivation began in the Heian period, sometime between the late eighth century and early twelfth century, around the same time as the Vikings invaded Europe. The cherry blossoms were objects of aesthetic and literary appreciation by wealthy aristocrats. The first-known cultivated cherry in Japan was a weeping cherry, a form of *Edo-higan*. Aristocrats were enchanted by the way in which the thin, supple branches bent over towards the ground, giving the illusion of tears when the tree blossomed, and so they propagated this mutation by collecting seeds and planting them in their gardens.

In AD 812 the imperial family hosted a cherry-blossom viewing party for the first time, establishing a link with the cherry culture that continues to this day. The Japanese aristocracy, which sought to forge a national identity away from Chinese influence, celebrated cherries as their own special flower. At their annual *hanami* gatherings they wrote poems about the flower and about life, and then read them aloud. In *The Tale of Genji*, a literary masterpiece written by Murasaki Shikibu in the early eleventh century, cherries were portrayed as symbols of youth, love, romance and contentment, even as the novel's main characters lamented the flowers' ephemeral beauty.

In the twelfth century, meanwhile, Saigyō, a poet and Buddhist monk, composed many poems about the *Yama-zakura* wild cherries at Mount Yoshino, the birthplace of the mountain-worshipping *Shugendō* religion. There the principal image of the religion, a three-eyed blue-black deity called Zaō Gongen, was often carved from a cherry tree, which was considered sacred. Saigyō expressed his love of nature by praying that his days on Earth would end when the cherries were in full bloom:

Let me die
Underneath the blossoms
In the spring
Around the day
Of the full moon.

As Ingram learned more about the significance of cherries in Japanese society, he was most fascinated by the *Sakoku* period of the nation's history. As we have seen, these were the tranquil years between 1639 and 1853 when Japan was largely closed to outsiders. It was when most cherry varieties had been cultivated, largely as an unintended consequence of a plan by the military commander Ieyasu Tokugawa to consolidate his power after becoming shōgun in 1603. Tokugawa kept his 270 or so regional *daimyō* lords under his thumb by establishing a central form of government in Edo (present-day Tokyo) that lasted until 1868.

In 1635 the third Tokugawa shōgun, Iemitsu, made it compulsory for each *daimyō* to live in Edo every other year — a system called *Sankin-kōtai* — as a way to maintain control. When these *daimyō* returned to their domains, their wives and heirs had to remain in Edo, almost as hostages. This system meant that each *daimyō* had to maintain luxurious mansions in at least two locations.

With the Tokugawa shogunate ruling the country strictly, battles between the rival clans of each regional domain died out. Many *daimyō* replaced their preoccupation with warfare with other pursuits, including the creation of beautiful gardens in Edo containing cherry trees. The *daimyō* felt obliged to make spectacular displays because the shōgun occasionally visited their residences, and it was their duty to entertain him lavishly and to highlight their sophistication.

The *daimyō* lords brought flora to their Edo homes from their regional domains. From the north came Japanese roses and skunk cabbages. From central Honshu came azaleas, bush clovers and various bellflowers. From the warmer south came peonies, hydrangeas and wisteria.

The *daimyō* also brought seeds of their favourite cherries from their homelands, which their gardeners planted, in the hope that aesthetically pleasing offspring would take root. After years of repeated trial and error, these trees' seeds produced ever-finer specimens. Occasionally the gardeners also grafted trees by cutting a small branch, called a scion, from a parent tree and inserting it into the 'root stock' of another tree. When a graft was successful, the new tree would grow on top of and out of the root tree, effectively taking over the root and vascular system of the tree onto which it had been grafted. This grafting technology, which had been practised in Japan since at least the thirteenth century, is still the most popular method of propagation today.

In these ways, by both serendipity and science, diverse cherries blossomed throughout the Edo gardens. As the *Sakoku* decades passed, the cherries thrived in number, variety and quality. While there is no accurate record of the number of cherry-tree varieties that existed by the 1860s in Edo, experts estimate about 250.

Some *daimyō* in Edo developed large gardens dedicated solely to cherries, called *sakura-en*. One of the most spectacular in the late eighteenth century was created by Sadanobu Matsudaira, grandson of the eighth Tokugawa shōgun and *daimyō* of the Shirakawa region north of Edo. He retired in the 1790s to devote himself to literature and gardening. Along with wisteria, plums and lotus plants, Matsudaira collected and planted 142 cherry varieties in his garden, on the site of what later became Tokyo's largest fish market in Tsukiji. Cherry trees, Matsudaira said, have 'branches so gentle, flowers so delicate in shape and hues so simple that the total effect is perfect beyond belief'.

Another high-ranking *daimyō*, Seihō Ichihashi, planted 234 cherry trees in his Edo garden, according to an illustrated book published in 1803. Yet another, Ryōzan Hori, the lord of the Suzaka domain north of Edo, published a book about cherries in 1861 called *Jaku-fu* that contained paintings of 250 cherries from 182 varieties.

It wasn't only the elite who enjoyed the cherry pastime. During the Edo period of 1603–1868 thousands of trees were planted in public places for ordinary people. This led to cherry-blossom viewing, *hanami*, becoming a phenomenon for the masses, and not just the garden-owning aristocracy and the *daimyō* lords. For

Taihaku, painted by Collingwood Ingram

Taihaku at The Grange

Somei-yoshino in Mie Prefecture, Japan

Somei-yoshino in *Sakura Zuhu* by Manabu Miyoshi

Yamazakura in Mie Prefecture, Japan

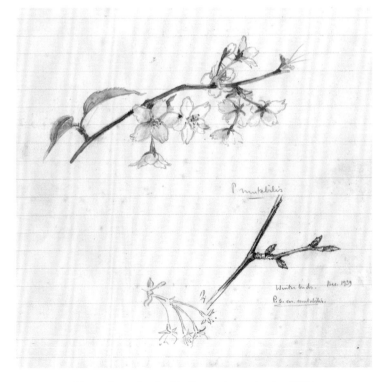

Yamazakura sketched in Ingram's notebooks, 1939

Kanhi-zakura, described as *Prunus campanulata,*
and a Sargent cherry leaf from Ingram's notebooks

Sargent cherry *(Ōyama-zakura)* in Nagano Prefecture, Japan

example, when the third shōgun, Iemitsu Tokugawa, built the Kaneiji temple in 1625 in Ueno, Tokyo, for the religious protection of Edo castle, hundreds of *Yama-zakura* from the Yoshino mountains were planted in the temple grounds. Several other *hanami* spots were created in Edo during the eighteenth century, where the general population caroused each spring under the trees.

11. The Dejima Doctors

Despite the sophistication of Japanese horticulture during the eighteenth and early nineteenth centuries, the world knew little about the ways in which trees and plants, particularly cherries, flourished in Japan until the *Sakoku* era was over. Yet as Ingram continued his research, the names of three foreigners kept cropping up. These were the trio of 'Dutch' doctors – Engelbert Kaempfer, Carl Thunberg and Philipp von Siebold, none of whom was actually Dutch – who had lived like prisoners on Dejima, the 600-foot × 240-foot man-made island off the city of Nagasaki in western Japan.

The third Tokugawa shōgun, Iemitsu, had commanded the island's construction in 1636 as a prison for Portuguese residents of Nagasaki, in order to prevent the spread of Christianity. Three years later all the Portuguese were expelled, and the fan-shaped island was leased in 1641 to a small group of men associated with the Dutch East India Company. For the next two centuries that island was Japan's only link with Europe and became a gateway, albeit limited, for Western culture, medicine and technology. That it was such an isolated and lonely place was probably why the Dutch authorities imported Japan's first billiards table for the fifteen or so inhabitants who lived there claustrophobically year-round.

On penalty of death, traders and visiting sailors were prohibited from crossing the well-guarded 40-foot stone bridge from Dejima to the port city of Nagasaki. Watching from afar in Edo, the shōguns had only one reason for allowing the Dutch their constrained presence, and that was trade. Twice a year four Dutch sailing ships arrived with a treasure trove of goods from elsewhere in Asia: sugar, cloves and other spices from Indonesia; deer skins from Tainan, the oldest city in Taiwan; and shark skins from Ayutthaya in Thailand. These were exchanged in Dejima's warehouses for silver and high-grade Japanese copper, which were then used in Western coins, ships, buildings and weapons.

The 'three scholars of Dejima', so-called because of their erudition, endured the island in different centuries: Kaempfer in 1690–92, Thunberg in 1775–6 and Siebold in 1823–9. All were highly skilled men of science and medicine, but were also

Dejima Island, Illustrated London News

fascinated by plants and trees. Because of their medical proficiency, they were each allowed to leave Dejima to accompany the *capitan* of the Dutch delegation on trips to Edo to meet the shōgun. In this way they became the first Westerners to collect and then spread the word about Japanese flora.

On a cool spring day in 2017, a year notable for the late blooming of the cherries, I sat quietly in the tiny Dejima garden where the 'Dutch' doctors had grown vegetables and flowers hundreds of years earlier. In one corner stands a stone monument that Siebold erected in 1826 to his predecessors, Kaempfer and Thunberg, in tribute to the garden they had established there. The Latin inscription, still visible today, reads: '*Ecce! Virent vestrae hic plantae florentque quotannis*' ('Behold! Your plants flourish and bloom here year upon year'). It was a magnanimous gesture to fellow scholars whom Siebold had never met.

In the 1820s, the island's one and only warehouse-lined street would have been a seething, unhygienic mass of sailors and traders, clerks and courtesans when ships were in port. Each ship carried 100 or so Dutch sailors. I thought about the excitement these visits must have brought to Nagasaki residents, as they tried to glimpse the tall, white-skinned Westerners while the provisions were offloaded from the ships.

Dejima was demolished after the Meiji government was set up in 1868, and the land around the island was reclaimed so that the Dutch settlement became part of Nagasaki. In recent years the government has started reconstructing the island. These days it's a spotlessly clean historical museum. But, alas, it's still part of the city, surrounded on all sides by noisy streets and high-rise buildings. The city's goal is to restore the buildings and stone walls, surrounded by a moat, to their original state by 2050.

In the 1920s, as Ingram read the doctors' accounts of their lives on Dejima, the most interesting passages would have been about the gruelling 1,600-mile round-trip journeys that the men had made, from Nagasaki to Edo and back, mostly on foot, to pay their respects to the Tokugawa shōgun. These ritual trips were the most important responsibility of the head of the Dutch contingent in Dejima, ensuring that the shōgun would continue to allow the lucrative trade.

Between 1633 and 1850 the various Dutch chiefs made 166 trips to Edo, accompanied by sixty-odd Japanese interpreters, cooks, guards and carriers of the *kago*, the wooden palanquin in which the leaders were transported. The men left Nagasaki in January and returned home in March, often meeting the shōgun for only a few minutes, despite the months of preparation and travel. For the Dutch chiefs, it was a necessary obligation. For the doctors, these journeys allowed them to gather information about unknown plants and trees, including cherries.

As Ingram noted, Engelbert Kaempfer was the first Westerner to mention Japanese cherries in European literature. In 1692, after one of his two visits to Edo to meet the fifth Tokugawa shōgun, Tsunayoshi Tokugawa, Kaempfer wrote: '[At] the beginning of the spring, the trees — chiefly plum, cherry and apricot trees — are in full blossom and loaded with numberless white and incarnate flowers, single and double, and no less remarkable for their largeness and plenty than for their singular beauty.'

En route, the German-born Kaempfer stayed in Japanese inns, each of which was immediately nailed shut after the foreigners arrived, 'to prevent any thoughts of an escape'. The only respite was a square garden at the back of each lodge, where a plum, cherry or apricot tree was always growing. 'The older, the more crooked and monstrous this tree is, the greater value they put upon it,' Kaempfer wrote. 'The great number of beautiful, incarnate and double flowers which they bear ... are a surprisingly curious ornament, but they have this disadvantage that they bear no fruit.'

Carl Thunberg, a Swede who arrived in Japan in 1775, eighty-three years after Kaempfer's departure, was more of a naturalist than his predecessor. He had been taught at Uppsala University by the renowned plant-classification scientist Carl Linnaeus, who had encouraged him to see the world. Thunberg took that advice to heart, travelling with the Dutch East India Company as a doctor, first to South Africa and then to Dejima, where he became head surgeon of the trading post. There he ingratiated himself with his interpreters and then with Nagasaki doctors who were looking for a cure for syphilis, known as the 'Dutch disease'. Among the few Japanese visitors allowed onto the island of Dejima were courtesans from Nagasaki's entertainment district, who received bags of sugar for their services.

The Japanese authorities allowed Thunberg, like his fellow doctors, to leave Dejima occasionally, culminating in a visit to the shōgun in 1776. Along the way, and on his one-day excursions near Nagasaki, he observed and collected numerous plant specimens. These formed the basis for his 1784 book, *Flora Japonica*, which was first written and published in Latin.

But it was the botanical research of the third Dejima doctor-botanist, Philipp von Siebold, that most interested

Ingram. He called Siebold's multi-volume German-language book, *Nippon*, a 'magnificent work'. Born in Würzburg in Bavaria in 1796, Siebold had arrived in Japan in August 1823, by which point it had been a closed nation for more than a century. Japanese scientists were by now impatient to learn about Western medical techniques and advances, such as smallpox vaccinations. Some scientific books had found their way into the hands of the Japanese elite after the eighth Tokugawa shōgun, Yoshimune Tokugawa, relaxed a ban on some Western literature after 1720.

A Japanese portrait of Philipp von Siebold

But there was still an immense knowledge gap, and virtually no one in Japan would have known about the seditionist ideas of liberty and progress that dominated intellectual discussion in Europe's capitals during the eighteenth century.

In the 1820s, with the permission of the authorities, Siebold started the Narutaki Clinic and School in the hills north-east of Nagasaki, about three miles from Dejima. For the Japanese, this was a groundbreaking experiment. Siebold's lectures on medicine and other aspects of Western learning, known as *rangaku* (Dutch studies), drew physicians and students from all over Japan. In return, Siebold asked his students to help him with his other interests, Japanese fauna and flora.

The authorities allowed him to share a home next to the clinic with a Japanese woman, Otaki Kusumoto. There, in a garden behind the house, Siebold grew hundreds of native plants. He later took many of these back to Europe – including, it is thought, the Japanese knotweed, an invasive weed that has ruined many a European garden.

On a visit to Edo in early 1826, Siebold had fascinated the doctors of the eleventh shōgun, Ienari Tokugawa, and the city's intellectuals by dissecting pigs' eyes, conducting cleft-lip operations on newborn babies and giving smallpox vaccinations to children. All were revolutionary procedures to the Japanese. But the doctor's time in Japan was cut short after he obtained detailed maps of the country, which the government, fearful of a foreign invasion, considered strategically important.

So it was that in 1828, five years after Siebold's arrival in Japan, he was accused of high treason and of spying for Russia. Placed under house arrest, he was expelled from Japan in December 1829, leaving Otaki and their two-year-old daughter, Ine. While the authorities confiscated part of his collection of plants, books, maps, woodblock prints and other artefacts, Siebold returned to the Netherlands with 120 crates full of materials. There he published widely, notably *Flora Japonica*, begun in 1835, which described more than 100 Japanese plants, ranging from camellia and the mophead hydrangea to the rhododendron.

Thirty years after Siebold's expulsion, the Japanese government lifted his ban and in 1859, at the age of sixty-three, he was finally able to return. It was an unhappy homecoming and he left Japan within three years. However, during his long absence his daughter, Ine Kusumoto, had become Japan's first female obstetrician. By chance, Ine was practising obstetrics in Nagasaki at the same time as my great-great-grandfather, Gon Abe, was studying Western medicine there.

Gon Abe had moved to Nagasaki in the late 1860s from Yonago, a city on the north coast of central Honshu, several years after Siebold's final departure from Japan. He was one of the progressive Japanese doctors who was eager to learn about Western medicine. Did Gon and Ine meet? In the rarefied world of Nagasaki-based Occidental physicians, there is a high probability that they did.

12. Hunting Plants

The plant-hunting activities of the Dejima doctors shed some light on Japanese cherry culture before Japan opened its doors to the outside world. But after the 1850s many more accounts were published in the West, often by the succession of British, European and American plant-hunters who tramped through Japan in the second half of the nineteenth century. Most gathered plants for display and sale back home, ranging across the spectrum from azaleas and chrysanthemums to magnolias and violets. For example, Exeter-born John Gould Veitch and Robert Fortune from Berwickshire in Scotland both arrived in 1860 and competed to see how many plant and other specimens they could take back to the UK.

Among Fortune's favourite exports from Japan were the pink clusters of the double-blossomed cherry variety known as *Takasago*, which was among the first Japanese cherries to reach England, in 1864. 'All countries are beautiful in spring, but Japan is pre-eminently so,' Fortune wrote in a memoir published in 1863. 'The double-blossomed cherry trees and flowering peaches were most beautiful objects, loaded as they are now with flowers as large as little roses.'

The early collectors were followed by botanists such as Charles Maries, who first visited Japan in 1877 and introduced more than 500 plant species to England from Japan, China and India. Later, John Gould Veitch's son, James Herbert, also a distinguished horticulturalist, joined Charles Sprague Sargent, the first director of the Arnold Arboretum at Harvard University, on a plant-hunting expedition in 1892. Sargent's interest was in comparing Japanese varieties of elm, maple, poplar, walnut, willow and witch-hazel trees with their cousins in North America.

Although most of these early Western visitors to Japan were men, a handful of adventurous women made their mark. One was a friend of Charles Darwin, the botanical artist Marianne North, who visited Japan in 1875 and painted numerous little-known plant varieties. Another was the American journalist and travel writer Eliza Scidmore, who first visited Japan in 1884 and later fell in love with the cherry-blossom varieties that lined the Arakawa River in Tokyo, once one of the most popular cherry-viewing spots in Japan.

Yet it was the more visual descriptions of cherry blossoms by the journalist Lafcadio Hearn that really caught the public imagination in the West and helped to wipe out the earlier prejudice that cherries should bear edible fruit:

> And I see before me ... a grove of cherry trees covered with something unutterably beautiful – a dazzling mist of snowy blossoms clinging like summer cloud-fleece about every branch and twig; and the ground beneath them, and the path before me, is white with the soft, thick odorous snow of fallen petals.

> Why should the trees be so lovely in Japan? With us, a plum or cherry tree in flower is not an astonishing sight; but here it is a miracle of beauty so bewildering that, however much you have previously read about it, the real spectacle strikes you dumb. You see no leaves – only one great filmy mist of petals.

Another likely source for Ingram was *Things Japanese*, a dictionary about Japanese culture, first published in 1890. The author was Basil Hall Chamberlain, a Professor of Japanese at Tokyo Imperial University, who was lauded for introducing Japan to the West. 'Old Japan is dead and gone, and Young Japan reigns in its stead,' he wrote. '[But] it is abundantly clear that more of the past has been retained than has been let go. The national character persists intact.' To make his point, Chamberlain quoted two poems that Japanese nationalists of the late nineteenth century frequently cited as being representative of the Japanese character. The first was by an eighteenth-century classical Japanese scholar, Norinaga Motoori:

> If someone asked
> What is the spirit of a true Japanese?
> I would say it is the Yama-zakura blossoms
> Shining in the morning sun.

The second was a popular Japanese proverb from the seventeenth century:

> The cherry is first among flowers,
> As the samurai is first among men.

Ten years after Chamberlain's cultural dictionary, a Japanese educator published an English-language best-seller called *Bushidō: The Soul of Japan*, about samurai ethics and Japanese culture. Japan's victory in the 1894–5 Sino-Japanese War had stunned the West. The book's author, Inazō Nitobe, wanted to help Westerners understand the source of Japan's strength at a time when regional dominance was shifting to Japan from China. Nitobe had become a Quaker while studying at Johns Hopkins University in Baltimore. There, he met and married an American, Mary Elkinton.

At the heart of Japan's national character, Nitobe said, was an unwritten code of moral principles that was passed down the generations, and which the samurai were required to obey. US President Theodore Roosevelt read and distributed the book to his friends around the time of Japan's victory over Russia in the Russo-Japanese War of 1904–5. *Bushidō* explained the virtues of chivalry, filial piety, loyalty, patriotism, politeness and self-control. Once the preserve solely of the samurai, the 'spirit of *bushidō* has permeated all social classes, furnishing a moral standard for the whole people,' Nitobe wrote. The cherry blossom, he added, was both 'the favourite of our people and the emblem of our character'. And he lambasted the English rose for its hidden thorns, showy colours, heavy scent and the tenacity with which it clung to life:

> All these are traits so unlike our flower, which carries no dagger or poison under its beauty, which is ever ready to depart life at the call of nature, whose colours are never gorgeous, and whose light fragrance never palls. Beauty of colour and of form is limited in its showing; it is a fixed quality of existence, whereas fragrance is volatile, ethereal as the breathing of life.

To Ingram, a romantic who had grown up reading tales of Arthurian chivalry and honour, the samurai virtues that were enshrined in the *bushidō* philosophy seemed pleasantly familiar. In all likelihood, he first read about the samurai spirit in *Tales of Old Japan*, the book by Algernon Freeman-Mitford, who had been stationed in Japan during the Meiji Restoration. His book, first published in 1871, popularised unknown Japanese stories, fairy tales and legends that fascinated Westerners:

> The feudal system has passed away like a dissolving view before the eyes of those who have lived in Japan during the last few years. Nor have heroism, chivalry and devotion gone out of the land altogether. We may deplore or inveigh against the Yamato Damashi, or Spirit of Old Japan, which still breathes in the soul of the samurai, but

we cannot withhold our admiration from the self-sacrifices which men will still make for the love of their country.

Freeman-Mitford, the paternal grandfather of the Mitford sisters, was influenced, as was Ingram, by the Irish gardening author William Robinson. In 1886 Freeman-Mitford inherited the Batsford estate in the Cotswolds from a cousin. After moving there in 1892, he created a wild Oriental garden filled with Japanese and Chinese artefacts and with masses of bamboo.

As Ingram continued his research, it became clear to him that the blossoms of cherry trees in Japan were far more than mere flowers. Indeed their role in Japanese society was far greater than that of any other national plant. At two well-attended cultural events – the Exposition Universelle, or world fair, in Paris in 1900, and the Japan–British Exhibition of 1910 – the Japanese government proudly displayed cherry trees in front of crowds of fascinated Europeans.

The Paris fair included a Japanese section where cherries were on show alongside *bonsai* (cultivated miniature trees) and chrysanthemums. The London event was more grandiose. Japan's Harvard University-educated foreign minister, Jutarō Komura, who had earlier been ambassador to Great Britain, was eager to show off his nation's power and culture. In the grounds of London's White City, Japan constructed a five-acre garden in which they planted hundreds of trees and shrubs that originated in Japan, including cherry trees procured in England. More than eight million people visited the six-month-long exhibition, entering the grounds under two artificial cherry trees flanked by samurai warriors.

Japan's 'cherry blossom diplomacy' also spread to the United States. Many American plantsmen had shown greater early interest in Japanese flowering cherries than their contemporaries in Europe. This prompted the Yokohama Nursery Company, a pioneering horticultural trading firm, to open a branch in San Francisco in 1890 and a second in New York in 1898. In addition to the Arnold Arboretum's director, Charles Sprague Sargent, one other early cherry enthusiast in America was David Fairchild, a botanist at the US Department of Agriculture, who visited Japan in 1902.

Fairchild and his wife, Marian, the daughter of the inventor of the telephone, Alexander Graham Bell, imported twenty-five cherry varieties into the US in 1906 and planted them at their Maryland home, to see how well they would grow. Astonished at the cherries' beauty, Fairchild ordered 150 weeping cherries from Yokohama Nursery and arranged for each school in Washington D.C. to plant one on Arbor Day in 1908. Fairchild also broached the idea of planting thousands of cherries in the capital's Potomac Park. This suggestion delighted Eliza Scidmore, who was a prodigious writer for the National Geographic Society's magazine and was the society's first female trustee. She seized on Fairchild's proposal and wrote a letter in April 1909 to the First Lady, Helen Herron Taft, proposing that cherries should be planted along the Potomac River.

When the newly elected president's wife accepted the plan, Tokyo mayor Yukio Ozaki sent 2,000 cherry trees to Washington D.C. in November 1909 to show his gratitude for America's role as a go-between during the Russo-Japanese War. This first shipment of cherry trees proved to be infested with insects, and American plant-quarantine officials had to incinerate them. A second set of trees was sent to New York, which was celebrating the tricentenary of the discovery of the Hudson River; they failed to arrive because the steamer carrying them sank en route.

Undaunted, Mayor Ozaki arranged to donate 6,020 cherry saplings in 1912. After arriving in Seattle from Yokohama, half the shipment travelled in refrigerated trucks to New York and the other half to Washington, where Scidmore watched President Taft's wife plant the first tree, next to the Tidal Basin. Today the National Cherry Blossom Festival is among the most spectacular and popular events each spring in the US capital, attracting one and a half million visitors annually. Meanwhile the saplings sent to New York were spread throughout the city. Some were planted in Claremont Park on Manhattan's Upper West Side, which was renamed Sakura Park; others in Central Park and along the banks of the Hudson River.

During the 1910s the atmosphere was ripe for the introduction and dissemination of cherry blossoms on both sides of the Atlantic. But when the First World War began, cherries that bore no edible fruit took a back seat to each nation's war effort. It wasn't until the early 1920s that several wealthy cherry aficionados in Europe once again started importing, planting and collecting the trees. In this endeavour no one was more enthusiastic and committed than one of Benenden's newest residents, Captain Collingwood Ingram.

13. Creation and Collection

From the wooden gates at the bottom of the driveway to The Grange, Collingwood Ingram and his gardener, Sidney Lock, surveyed their work. On either side of the curving carriageway they had planted cherry saplings in the clay soil. To protect each shallow-rooted tree from storms, the men had driven a stake into the ground as a support beside each tree, attached with pieces of old bicycle tyres. Between the trees they had planted rhododendrons, azaleas and other medium-sized

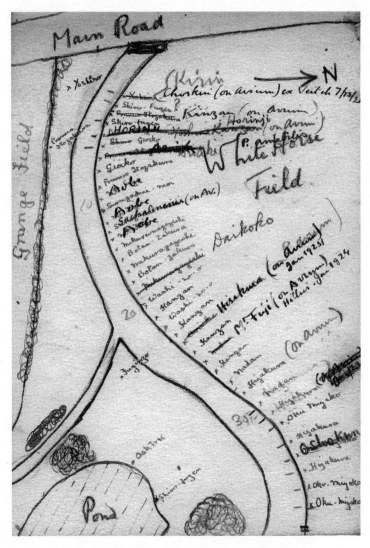

Ingram's plan of the cherry planting at The Grange

shrubs. Behind these saplings, they used shovels and picks to dig out a garden from the meadow behind the house.

The composition and structure of the garden reflected numerous global influences. The priority was to produce the most natural garden possible, one in which there was neither precise symmetry nor any perfectly straight lines. 'My aim was to reduce to an absolute minimum the unavoidable artificiality of a man-made pleasance,' he wrote.

In the meadow, Ingram planted Scotch pines about 20 feet apart, as a wind-resistant perimeter hedge, with rhododendrons in between. He had bought every possible cherry-tree cultivar that he could find, ranging from the late-blossoming *Fugenzō*, one of the oldest varieties in cultivation, to *Ukon*, which was known for its greenish-yellow semi-double blossoms.

One excellent source of plants was the Yokohama Nursery, which had opened a London office in 1907 to import exotic Japanese plants. The company still exists today. Its early English-language catalogues, stored in the Yokohama Kaikō Museum, highlight some of the most popular cherry varieties on sale in the early twentieth century.

As well as buying commercially imported plants, Ingram also tapped his burgeoning network of cherry friends around the world to send or swap varieties. Some came from the Royal Botanic Gardens in Kew, for instance, where he was friends with William Jackson Bean, the Yorkshire-born curator of the gardens during the 1920s. During the 1900s Bean was the first to plant Japanese flowering cherries at Kew, including the *Temari* variety and Sargent cherry, brought by Charles Sprague Sargent from Hokkaido in 1892. The British plant-hunter and cherry expert Ernest Henry Wilson (usually known as E.H. Wilson), who had worked at Kew in 1897, had also sent some cherry seeds to the gardens, which by the time of Bean's tenure had become blossoming trees.

Elsewhere, Ingram obtained seeds from Colonel Robert Stephenson Clarke, who owned a 200-acre estate at Borde Hill, Sussex. By financing plant-hunting expeditions to the Himalayas and China, Clarke had accumulated one of the world's largest private collections of trees and shrubs. Meanwhile, at the Royal Botanic Garden in Edinburgh, Ingram dealt with one of Scotland's pre-eminent plant-hunters, George Forrest, who had led numerous expeditions to China. From the Donard Nursery Company in Northern Ireland's County Down he received in 1923 the hardy and disease-resistant Fuji cherry, another of Japan's ten wild species.

Many others came from the United States. Ingram pleaded with Arnold Arboretum to send him seeds or scions of the plants that he didn't have. The arboretum contained many cherries that E.H. Wilson had collected on its behalf in Japan in 1914 and later.

Ingram obtained other cherries from W.B. Clarke & Co., a nursery founded by Walter Bosworth Clarke in San Jose, California. Still more came from Clarence McKenzie Lewis, a wealthy widower and trustee of the New York Botanical Garden, who in 1922 had bought a 1,000-acre estate in northern New Jersey and was building a timbered house there, called Skylands Manor. It was surrounded by gardens containing plants from around the world, including rare cherry varieties. Through these and other contacts, Ingram established a transatlantic 'cherry network' that kept him well informed about any projects involving cherries in the continental United States.

When Ingram obtained a scion, he grafted the branch onto another tree, usually a native British *Prunus avium*, a wild cherry species. Grafting wasn't always successful, but over time he propagated scores of varieties at The Grange, and his garden boasted a diverse and unique collection of wild-cherry species and cultivated varieties from around the world. By 1925, barely six years after moving to Benenden, some seventy cherry varieties were already growing in his garden, and Ingram knew more about cherry trees than all but a handful of professional gardeners in the West.

14. The Hokusai Connection

When Ingram wasn't in his garden or travelling, he was usually holed up in his untidy attic, researching and writing. His first notable paper about cherries, titled appropriately 'Notes on Japanese Cherries', was published in the prestigious *Royal Horticultural Journal* in 1925. Throughout the 1920s, on long winter nights in his hideaway, Ingram became a cherry-sleuth, sorting through piles of dusty research papers, nursery catalogues and his own notes to find clues about each plant's identity and name. It was a tedious and frustrating task.

When British nurseries started importing cherries from Japan and the US at the turn of the century, they had often given them English names to make them more appealing and marketable. The sweet-smelling, light-pink blossoms of the *Amanogawa* variety, for instance, looked like apple blossoms. Hence, they were called 'apple blossom' cherries, even though the translation of *Amanogawa* – 'Milky Way' or 'Heavenly River' – was equally descriptive. Another cherry, the white *Shirotae*, was sold as the 'Mount Fuji cherry'. It was a classification nightmare.

Many specific varieties had numerous names. For example, the *Kanzan* variety, which was particularly popular in Britain, was known in different parts of Japan as *Sekiyama*. It was also called 'New Red'. Other groups of varieties had been bundled into one name, even though it was clear to Ingram that they had dissimilar characteristics. 'In attempting to create order out of the chaotic confusion of cherry names that exist in Japan and Europe, I find I have set myself an almost impossible task,' he wrote. 'My greatest difficulty is the pernicious practice in vogue among European and American nurserymen of inventing their own names. It is very discouraging to buy the same plant, as I have done, under a dozen different names. The Japanese themselves are often equally to blame.'

Ingram asked two Japanese cherry experts to help unravel the mysteries of the cherries' names. One was Manabu Miyoshi, Professor of Botany at Tokyo Imperial University, who was known as the 'cherry professor'. The other was Gen-ichi Koizumi, a Kyoto Imperial University professor, who in 1913 had published a paper on roses. This included information about cherry trees, given their place in the rose family. While the cherry tree's history stretched back more than 2,000 years, scientific study and official categorisation of the numerous varieties had not begun until the early twentieth century.

Miyoshi, who had studied botanical physiology at Leipzig University in Germany, was one of the fathers of Japanese botanical studies, with a special interest in wild-cherry species and their cultivated varieties. In 1916 he had published a groundbreaking academic paper in German that summarised for the first time the taxonomy of more than 100 cultivated varieties. It was based on observations and research conducted on the banks of the Arakawa River.

As Ingram attempted to understand the cherries' heritage, a third source of information was E.H. Wilson's *The Cherries of Japan*, published twenty days after Miyoshi's German-language paper. As great rivals, Wilson and Miyoshi had each been determined to publish a comprehensive cherry guide first, and neither mentioned the other in their books. In the end, when Ingram was stumped about a variety, he simply named it himself. 'I do not suggest for a moment that these are really new varieties, but it is better to rechristen a plant than to run the risk of bestowing upon it a name already occupied,' he wrote.

Ingram was particularly interested in deducing the heritage of the two tall and vigorous trees in his garden that had first sparked his interest in cherries, on arrival at The Grange. He sent some leaf samples to Professor Miyoshi, who replied that the cherry did not have a name. And so Ingram made one up. He called the variety *Hokusai*, after the world-famous Japanese woodblock painter whose work he loved.

There was an affinity here, for both men were bewitched by the beauty of Mount Fuji. Ingram called it 'the world's most shapely mountain' and said the volcano possessed an 'almost supernatural grandeur' that seemed 'to belong to another world – to a new and unattainable cosmos, a materialized cloudland'.

Hokusai, a member of Buddhism's Nichiren sect, believed that Japan's tallest mountain possessed secrets of immortality. Mount Fuji and cherry blossoms appeared in his most famous

Fuji from Goten-yama, on the Tōkaidō Highway, *Katsushika Hokusai*

woodblock prints, which were usually carved on cherry wood from the mountains.

Ingram, himself an accomplished artist of birds, animals and nature themes, perhaps also felt a kindred spirit with Hokusai, though separated by time and place. The two men were born 120 years apart, almost to the day – Hokusai on or around 31 October 1760, Ingram on 30 October 1880.

As for the *Hokusai* tree, it was a hardy, long-lived variety with an 'excellent constitution', Ingram wrote. One of the two in his garden was the largest-known *Hokusai* tree in the world: 25 feet high and with a 42-foot branch span. 'When, in the spring, every branch is literally smothered in pale-pink blossom, it would be difficult to conceive a more striking floral display.'

Despite Ingram's achievements in the early 1920s, he was not satisfied. Japan was the Holy Grail for cherry-lovers and it was now almost two decades since his previous trip. Frustrated that he couldn't find more varieties in Britain, he concluded that the only way to become a real cherry expert and expand his collection was by making another visit to Japan.

This growing conviction was galvanised by a visit to The Grange from Japan's so-called 'bird prince', Nobusuke Takatsukasa, at the peak of the 1924 cherry-blossom season. A wealthy ornithologist with connections to the Japanese royal family, Duke Takatsukasa's association with Ingram stemmed from his studies in Tokyo University's zoology department, under Professor Isao Iijima. On Ingram's 1907 honeymoon visit, Iijima had helped him obtain bird-catching permits. Like Ingram in his early years, Takatsukasa was obsessed with birds, and he had come to Europe to pursue his research. The duke's visit to Benenden was to change both men's lives. Takatsukasa was particularly impressed with the *Yama-zakura* trees, the most popular wild-cherry species, that were growing at The Grange. '*Yama-zakura* in England were so beautiful that it made me feel I was back in Japan,' he said later. Ingram had imported the species in 1920 and believed they were the first to reach Britain.

With advice from the duke and his influential connections, Ingram drew up a wish list of four unmissable places where he hoped to collect varieties he did not already possess. First, the temples and shrines of Japan's ancient capital of Kyoto, where cherry varieties had thrived for over a thousand years. Second, Tokyo, and in particular the Arakawa riverbank and the western district of Koganei. At the former, more than 3,000 trees of seventy-eight cultivated varieties had flourished since 1886; at the latter, Japan's eighth shōgun had planted thousands of *Yama-zakura* trees in the 1730s and 1740s. Third on Ingram's list were the foothills of Mount Fuji, where wild cherries grew in abundance. Lastly he wanted to visit Nikko, a small city in the mountains 100 miles north of Tokyo, where Ieyasu Tokugawa, founder of the Tokugawa shogunate that ruled Japan from 1603 to 1868, was buried, and where azaleas, cherries and hydrangeas thrived in the cool air. It was an ambitious itinerary, but by a combination of foot, horse, motor boat, steam train, car and ship, Ingram was determined to see Japan's top cherry sights at the peak of the blossoming season.

Duke Takatsukasa also arranged meetings for Ingram with Japan's three leading cherry experts: Seisaku Funatsu, the 68-year-old doyen of Japanese cherry trees; 44-year-old Count Tsuneo Kajūji, Kyoto's foremost authority; and 65-year-old Manabu Miyoshi, the 'cherry professor' who had already helped Ingram with classification issues. The duke also set up meetings for Ingram with members of the Cherry Association, an elite organisation set up to promote and protect cherries, as well as owners of nurseries and gardening experts. Ingram tacked on a visit to the southern island of Kyushu, where he intended to collect other plants, such as azaleas and hydrangeas.

After months of careful planning, mostly by Duke Takatsukasa, Ingram boarded a ship to Singapore, spent a couple of weeks en route in Sumatra, and headed for Japan at the end of winter 1926.

P. incisa v. alpina

PART THREE

SAVING THE
SAKURA

ap./41

note long style a.

Flowers white about 2 cm across
Sessile umbel - 2-3 flowers.
Pedicel about 1.2 cm long - fruit
thick - sparsely hairy
Sepals were - red.

Serration, double or
treble, v. deeply cut

leaf up to 7c. long
(mostly much less)
& 3.2 broad.

— Petiole redd
glandular
about 1 c
(glabrou

Leaf rather rugose.
Sparsely & shortly bristly above.
Sparsely bristly on mid-rib & main veins

young wood golden brown.   7 to 8 pairs of veins.
                              Stipule lanceolate: toothed - up to 8 mm long.

10.5.43

Alpine form of P. incisa (P. alpina) ex Wale

Leaves at first hairy on under surface, especially on midrib. Acuminate tip. Serrate v
bi-serrate.

[up to 14 mm]
Total length of style about 12 mm. Cupula (glabrous) purplish-red, about 6 mm long by about
3.5 mm broad - more or less tubular with very slight swelling base. Sepals ∧ triangular
ovate. Pedicels, v. sparsely hairy about 1 cm long. Peduncle wanting. Leafy bract at base very
minute. Stamens about 21 - approx 3/5 as long as visible portion of style.

15. The Pilgrimage

Collingwood Ingram arrived in Nagasaki on 30 March 1926, in what would be the final year of Emperor Taishō's reign. He left about seven weeks later, 'veiled', as he wrote, 'in tears'. He was never to return to Japan. His visit was essentially a pilgrimage – a *sakura angya* – that he believed would broaden and deepen his knowledge of cherry culture to a greater degree than all but a few other people in the world. *Angya* is a Zen Buddhist term that refers to the walking pilgrimage that monks and nuns make as they prepare to become spiritual masters and guides.

This was a bittersweet visit. Not only did Ingram collect new cherry varieties for his garden, but he also visited the most famous cherry locations, met the nation's top cherry experts and was treated like royalty. And yet, two decades of yearning for a country that, in early adulthood, had taken his breath away evaporated within days.

Ingram's first impressions were alarming. On 1 April, when his ship steamed through the Inland Sea towards Tokyo, the mountaintops on the island of Shikoku, south of Honshu, were still smothered in snow. Few cherry trees were in full bloom, having been held back by the unusually long winter. That couldn't be helped, of course, but Ingram's second impressions were even more disheartening. On Friday, 2 April he travelled from Tokyo to Yokohama, 25 miles south of the capital, to meet the heads of the Yokohama Nursery, one of the main cherry-tree exporters. En route, he witnessed what he described as 'a litter of untidiness' that confirmed the disdain he had felt on his previous visits for urban Japan and industrialisation. He wrote in his diary:

> The old Oriental towns have been wiped out, and upon their sites are being reared ultra-Occidental buildings of great size and hideousness. It seems to me that Japan is trying to swallow too suddenly and at a gulp, a much too big dose of Western civilization, and that she is suffering in consequence from a sort of violent aesthetic indigestion.

In part, the events of 1 September 1923 were to blame. Two minutes before noon on that day the ground beneath Tokyo and Yokohama shook violently for ten minutes, triggering massive fires in and around the conurbation. This was the Great Kantō Earthquake, one of Japan's worst natural disasters. It caused a tsunami with 30-foot-high waves to roll in over the coastal villages. A typhoon barrelled through the capital the same day, its powerful winds fanning the flames. This combination of the 7.9-magnitude quake and its merciless consequences left more than 142,000 people dead, three million injured and 1.9 million homeless. To rebuild the metropolis, where more than 500,000 homes had been destroyed, the government pulled down thousands of wooden houses and replaced many with concrete buildings.

The result wasn't pretty. Even before the earthquake, the densely populated region had suffered from Japan's unrelenting push for economic growth and international recognition. The downside to the smoke-belching factories that had turned Japan into the world's ninth-largest economy was that they had also turned many cities grimy-grey.

Much like the pea-souper that had affected London at the time of Ingram's conception, the cities' smog was so bad that some children believed the natural colour of tree leaves to be grey rather than green. Moreover, the cities had become increasingly crowded, as jobs in the countryside diminished and urban employment soared. This had increased demand for company dormitories and hastily built apartments, at the expense of green spaces. After the disaster, business slumped, tens of thousands of workers were laid off, labour unions proliferated and the economy became chronically depressed.

The Yokohama Nursery lay in the heart of the area most affected. There, at the company's headquarters, Ingram found himself sitting on a sofa and drinking green tea as an interpreter translated the words of Kiyoshi Suzuki and Masunosuke Shimamura, respectively the nursery's president and general manager.

It was a dispiriting conversation. 'It appears,' Ingram wrote later, 'that the commercialization of Japan has caused the cult of these beautiful trees to wane.' Suzuki and Shimamura told him that although the Japanese people were still devoted to cherry blossoms, they no longer showed any interest in the varieties. When ordering a tree, they said, people merely asked for a 'single'- or 'double'-blossomed flower, and they seldom received any orders for a specific named variety.

Ingram was startled. For the first time he realised how much damage the dramatic shift from a feudal to a modern society was inflicting on the cultivated cherries. During the *Sakoku* era, the *daimyō* lords' gardeners had poured their time and energy into creating more beautiful and attractive varieties. After the Meiji Restoration, as the *daimyō* lost their social status and were pensioned off, many of their luxurious mansions and cherry-filled gardens in Tokyo were abandoned or fell into disrepair. Other gardens were turned into tea and mulberry plantations. Many cultivated cherries that the *daimyō* had tended for decades were cut down or died of neglect. And because each variety was unique to a specific *daimyō*'s garden, countless varieties became extinct.

To be sure, wild cherries continued to thrive in the mountains. However, in the unprecedented rush to revamp the economy, the Japanese had forgotten about the diverse range of cherries that had been developed during the Edo period. The one exception was the newly cultivated *Somei-yoshino* cherry, with its soft-pink blooms. Japan's new leaders were eager to find symbols of national unity and modernity with which the population could identify. The cloned cherry, developed in the 1860s, during the final years of the Edo period, fitted the bill perfectly.

Year by year, cherry trees in Japan were becoming less diverse. One of Ingram's priorities was to find new varieties there that were unknown to expert botanists. Yet only five days into his trip, the realisation of the cherries' impoverished condition hit him hard and raised lots of questions: Why had the tradition of creating cherries collapsed so completely after the Meiji era began in 1868? Was it possible to stir up demand for varieties again? If so, what would it take to revive interest in them?

Faced with these questions at the Yokohama Nursery, Suzuki and Shimamura had few answers, and the lengthy conversation left Ingram feeling disheartened, as his diary entry reveals:

We owe the extraordinary richness of varieties to bygone days, and especially to the Tokugawa period. Horticulture was then at its zenith and the cherries subject to critical selection. Now that no interest is taken to save the less-showy or (what is more serious) the less-easily propagated varieties from extinction, their numbers are sure to decrease.

But later that night Ingram reassessed the situation more positively:

> Happily, they [cherry trees] are not very short-lived plants and it may not yet be too late to save some from oblivion. [However,] in years to come, the Japanese will have to seek some of their best sorts [of cherries] in Europe and America.

As Ingram noted in his diary, he had at least four cherry varieties growing at The Grange that appeared neither in the Yokohama Nursery sales catalogue, which listed seventy-two varieties, nor in Professor Miyoshi's classification guide of 133 species and varieties. Already he was thinking about how to save Japan's rarest flowers.

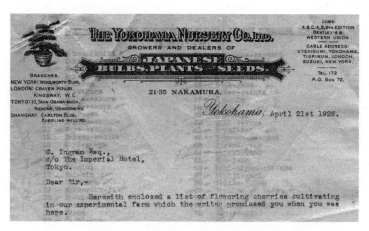

Ingram's 1926 correspondence with the Yokohama Nursery

16. Twin Pines

Ten days in Japan's ancient capital of Kyoto buoyed Ingram's spirits. His guide there was another plant-loving aristocrat, Count Tsuneo Kajūji, founder of the Kyoto Horticultural Club and the city's foremost cherry expert, and they were accompanied part of the time by Duke Takatsukasa. Ingram received royal treatment at the luxurious Miyako Hotel, where he was fawned over by the staff. Wherever he went, he couldn't escape what he called 'the interminable bowing and scraping'. Rather than join in, Ingram would quickly shake people's hands, 'mumble a few unintelligible words and assume an inane grin'.

All Ingram wanted was to see and collect cherries. Ambling through the gardens of Daigoji, a Shingon Buddhist temple, on 8 April 1926, he marvelled at the temple's collection of weeping cherries. One was even said to have been in blossom in 1598 when the de facto leader of Japan, Hideyoshi Toyotomi, held the nation's most lavish *hanami*, or cherry-blossom viewing, banquet there. To solidify his succession plans and flaunt his wealth, Toyotomi had rebuilt the temple, planted 700 cherry trees in a 'flower tunnel' and led a procession around the grounds with his wife, his concubines and his five-year-old son, Hideyori. A patron of the arts, Toyotomi immortalised the event in a poem:

> When flowers bloom
> Among twin pines
> The cherry blossoms

At Daigoji Temple

Will last forever.

The 'twin pines' were Toyotomi and his son. Yet the succession was doomed. Five months after the party Toyotomi was dead, and Japan descended once again into war. Five years later, in 1603, Ieyasu Tokugawa was declared shōgun. These days, however, Toyotomi's extravagant parade and party are re-created at the temple every April in the temple gardens, where more than 1,000 cherry trees now bloom.

When Ingram visited the temple, the sight of a weeping tree in full bloom, seen from indoors, had a powerful impact. 'With the strong sunlight striking through the cascades of soft pink blossoms, viewed through the open *shōji* [paper] doors, the effect was inexpressibly beautiful,' he wrote.

Finally Ingram was in *sakura* heaven, albeit still possessed by the need to find unknown or rare varieties. On 10 April he and Count Kajūji went to three historical Kyoto landmarks: the Kiyomizu Temple, founded in AD 780; the Kyoto Imperial Palace, the home of emperors before 1868; and Hirano Shrine, site of a cherry-blossom festival every year since AD 985.

It was in the Kiyomizu Temple grounds, close to a waterfall, that Ingram discovered a semi-double, large-flowered tree that he thought had been developed from an *Edo-higan* wild cherry. It was a cultivar that he didn't have at Benenden. Several boughs of the tree were dead, but Ingram was determined to preserve the specimen and asked the count to send him scions. A cherry scion that is grafted onto a root tree must be picked during its winter dormancy, which at this point in the spring was at least eight months away.

At the Kyoto Imperial Palace, meanwhile, Ingram noticed an unusual-looking *Yama-zakura* tree, the most common wild-cherry species, with extremely small flowers. Then at the Hirano Shrine he found three other cherries that he didn't own: a double-blossomed *Kiku-zakura*, a twin-fruit-bearing *Imose* and a soft shell-pink *Taoyame*. It was a day of delight.

The count promised to send scions of all these cherries to Ingram, along with scions of three other cherries that the visitor had seen: a weeping spring cherry from Maruyama Park, a renowned cherry-viewing park in Kyoto; a winter-flowering *Jūgatsu-zakura* from Nanzenji Temple, a Zen Buddhist temple established in 1291; and an unnamed cultivar with clusters of erect branches. In exchange, Ingram made a note in his diary to send English double-white lilacs to the count.

Ingram was still often confused by the names of these cherries. At the Hirano Shrine in Kyoto as many as fifteen varieties with which he thought he was familiar had unfamiliar names. 'Are these cherries named according to Professor Miyoshi's taxonomy system?' he asked one of the gardeners. 'No, sir,' the man replied. 'The professor uses names from eastern Japan [Tokyo]. Our cherries have western Japan [Kyoto] names.' Count Kajūji jumped into the conversation. 'Kyoto is the ancient capital, Ingram-san. There's no need to follow Tokyo customs.' It was a mammoth task to figure out which variety was which, because cherry experts and cherry sellers in different places were giving different names to the varieties.

But perhaps the greatest problem was one of perception. Most Japanese, including the government, simply weren't aware that the diversity of their cherries was decreasing. So there was no urgency to take action. After all, there were blossoms throughout

Ingram in Japan, 1926

the nation every spring, although many of the urban trees were of just one kind – the *Somei-yoshino* variety. It was also correctly assumed that the wild cherries and their offspring still grew in abundance in the mountains.

On the evening of 11 April 1926 Ingram met Kan Kōriba, one of Japan's most renowned botanists, and six of his male colleagues at the Miyako Hotel in Kyoto. Kōriba was head of the Kyoto Botanical Garden, which had opened two years earlier as a national showpiece. The Kyoto prefectural government had bought a 60-acre site on which it had planned an exposition in memory of the Taishō emperor's inauguration in 1912. But when it did not follow through on this plan, one of Japan's wealthiest and most influential families, the Mitsuis, stepped in, enabling the land to be transformed into the nation's first public botanical garden. Kyoto's intent was to rival Tokyo as the centre of culture in the 'new' Japan.

The conversation that evening was desultory because the six Japanese spoke little English, but Ingram noted a comment from one of the men, Mr Hamaguchi, about his favourite cherry. It was a comment that Ingram repeated throughout his life:

> 'We Japanese like the single white mountain cherry best because it resembles a simple country girl, with a strong, healthy peach-like complexion – the heart and spirit of the "real" Japan.' At the thought of these rural damsels, [Hamaguchi] waved his hands dramatically towards the mountains. 'They are not pretentious, not too brightly coloured, like the double ones.'

17. Cherry Meccas

Two dozen years earlier, on his first visit to Japan, Collingwood Ingram had enjoyed an exhilarating late-summer boat trip down the Hozu Rapids in western Kyoto. Now, at the height of spring 1926, the shallow boat streamed past a profusion of red camellias, pale lilac azaleas, squat masses of bamboo and drooping sprays of dazzling white spirea on the steeply wooded riverbanks. Finally, to Ingram's delight, masses of snowy cherry blossoms appeared, which 'seemed to glow like patches of light among the deep shadows of their sombre neighbours'. Yet even amid this beauty, Ingram was worried about the future of these wild *Yama-zakura* trees:

> This contrast of pure white against the translucent red-brown of the unfolding leaves gives the finest forms of this cherry so much beauty. On a wooded hillside, it appears to be a fairly long-lived tree. But uncared for, and uncultivated in the towns and cities, it does not make very fine specimens.

Ingram's fears were confirmed at Yoshino, a mountainous town south-east of the city of Osaka in central Japan, which for more than a thousand years had been a mecca for Japanese cherry-worshippers. Yoshino had become a popular destination in part because it stood on the forested pilgrimage trail through the Kii mountains in southern Honshu, which included many temples and shrines dedicated to the *Shugendō* religion, an esoteric mix of Buddhism, Shintoism and Taoism.

Close to this trail, thousands of cherry trees had been planted over the previous 1,400 years at four separate places and at different altitudes. In this way, because of the difference in climate at each site, the trees flowered at different times during the spring. However, in 1872 the Meiji government, determined to downplay Japan's feudal past, had banned *Shugendō* as a 'superstitious religion'.

At Yoshino, Ingram had expected to see dense masses of large wild *Yama-zakura* trees resplendent in their natural state. What he found instead was a motley group of thinly planted cherries and scraggy specimens. 'An apple orchard in the Alps is more beautiful,' he wrote. 'I confess I was more than a little disappointed with this much-vaunted *hanami* [flower-viewing] place.' Worse, Yoshino was overflowing with more than 20,000 visitors, half of whom were 'rollicking drunk on *sake*', the Japanese rice wine, Ingram wrote. The only saving grace was that the men were loud but happy drunks, becoming 'maudlinly affectionate with one another'.

It was time to depart for what Ingram hoped would be a more sober and enlightening locale – the Arakawa River in Tokyo – and to meet for the first time a man whom he called 'the fountain head of cherry lore': Seisaku Funatsu.

Funatsu, dressed in a dark-blue kimono and *geta* clogs, waited for Collingwood Ingram outside his small wooden home next to the Arakawa River, about eight miles east of the Imperial Palace. His traditional clothing, together with his flowing white beard and long fingernails, gave him an aura of Oriental authority and Confucian sagacity, as Ingram arrived to meet him on a bright, blustery day in late April 1926. The contrast between the two cherry-lovers could not have been greater. Ingram wore his usual white shirt, dark tie, light V-necked sweater, light suit and fedora hat, whether in a city or on a plant-hunting expedition.

Within Funatsu's house, Ingram recalled, the *shōji* doors were drawn back to let in the spring sunshine and the inevitable green tea and cakes were produced. Two other cherry aficionados joined them. One was Aisaku Hayashi, the University of Wisconsin-educated former general manager of the Imperial

Seisaku Funatsu, photographed by Ingram in 1926

Hotel from 1909 to 1922, who had created the influential Cherry Association in 1917 to protect and promote cherries; he acted as interpreter between Funatsu and Ingram. The other attendee was Yōichi Aikawa, a Tokyo city-parks official.

In the 1920s the banks of the Arakawa River were still one of the best places to view cherry blossoms in Japan, even though the number of healthy varieties had been dwindling for a decade. The Arakawa was Tokyo's longest river, stretching 107 miles from its source in the Chichibu mountains west of the capital to Tokyo Bay. The best viewing spot was a five-mile stretch of verdant riverbank where trees planted by local villagers had flourished since 1886, largely because of Funatsu's dedication.

The Arakawa blooms were a seasonal sensation, especially at the turn of the century. A myriad of cherry varieties produced blossoming colours that became known throughout Japan as the 'Five-Coloured Cherry Trees of Arakawa River'. Vast crowds arrived daily to see their various shades of white, pink, purple, red and yellow blossoms. As Professor Miyoshi, the botanical expert whom Ingram had met for the first time on 1 April, wrote:

> One could only marvel at the diversity of the colours
> and shapes of the Arakawa cherries. The trees were
> healthy, their branches stretching out fully, and plenty
> of flowers in bloom. Not only was each tree different
> in flower size, arrangement and tree shape, but some
> exuded fragrances. Suddenly, both the reputation of
> the cherries and the riverbanks were trumpeted far
> and wide.

Since 1903, when Professor Miyoshi first visited Arakawa, it was his habit to meet Funatsu at five o'clock each morning and jot down the distinctive characteristics and health of each tree, before the noisy throng arrived by steam train and boat. Funatsu then helped Miyoshi classify the many varieties for his monographs.

Walking along the riverbank, Ingram was struck by Funatsu's 'almost paternal interest in every tree' as he pointed out the beauty and merits of each variety:

Hanami *celebrations along the Arakawa River*

The love-light in his eyes was a joy to behold. I have often seen women looking with the same expression at their babies. He would stop and gaze rapturously up into the boughs. Sometimes he would use a pair of binoculars and stand for minutes, drinking in the exquisite beauty of the massed blossom.

As he stood there, the gusty wind thrashed the skirts of his dark blue haori coat about his sandaled feet and tossed his white beard to and fro like a banner. All the while, the pink petals kept falling softly around him, in little whirling eddies, like a shower of summer snow.

Some of the cherries were in superb condition, notably the *Ichiyō*, a double-petalled flower with large, soft-pink blossoms and young bronze leaves. Ingram also took note of the large-flowered semi-double *Shirotae*, the pale-pink *Benden* and the tiny white double-flowered *Kumagaya*, among others. But Funatsu told Ingram that many of the other varieties that had been planted there forty years earlier were in real trouble.

After heavy flooding in the Tokyo region in 1910, extensive drainage works were conducted along the Arakawa River, during which hundreds of these cherry trees were cut down. As pollution from the newly built factories and automobile exhaust fumes exacerbated the situation, the number of healthy varieties on the riverbanks began to dwindle. While Funatsu had repeatedly planted small trees as replacements, he had abandoned the initiative when these saplings were immediately stolen. The majority of the older trees were 'decaying or on the point of death,' Ingram wrote in his diary. 'Many of the varieties would undoubtedly have been lost to Japan ... were it not for his loving vigilance.'

18. Guardian of the Cherries

In fact it was somewhat of a miracle that many of the varieties growing on the Arakawa River still existed at all. That was largely thanks to two men: an obsessive cherry-collector called Takagi Magoemon and an equally passionate village mayor called Kengo Shimizu. Their dedication sparked an unlikely chain of events that ensured the cherries' survival in the mid-nineteenth century. While the Arakawa riverbank is today a shadow of its former self, the descendants of the cherries that once thrived there still blossom around the world.

Takagi Magoemon was the owner of the Baihō-en Nursery in Tokyo's Somei district and was the eleventh successive man in his family to inherit that name and position as a cherry expert. This was an illustrious pedigree stretching back to the seventeenth century, when the family provided plum and cherry trees to the ruling Tokugawa family. Since his great-grandfather's time, the Magoemon family had collected cherry varieties from the Edo gardens of some of the 270 *daimyō* lords, preserving them in the nursery.

The rare cherries in the *daimyō* gardens came under threat in the 1860s when the shogunate was weakened and many *daimyō* began returning to their home domains, abandoning their Edo gardens. Deeply concerned about the fate of these cherries, the eleventh Magoemon decided to save them himself.

One winter's day in the 1860s Magoemon visited a *daimyō* residence where he knew there was a spectacular double-blossomed cherry in the back garden. He knocked on the door and asked the *daimyō*'s wife to let him see the garden. Desperate to obtain a young branch to use as a scion, he took off one of his straw sandals, tied it to a piece of string and threw it over a thin branch. Tugging on the string, he bent the branch down sufficiently to snip the scion with scissors. Back home, he grafted it onto a root-stock tree and later propagated it.

Magoemon's pursuit of these unknown and disappearing cherries soon took over his life. Whenever his work schedule allowed, he would walk to a *daimyō* residence and tell the lord's wife that he wanted to preserve the cherry varieties in her garden and was collecting branches of the rarest trees. After *daimyō* families left Edo for the final time, Magoemon grew bolder. He clapped his hands together, murmured an apology and climbed over the fences of the gardens to gather more specimens.

As well as the scions he obtained, Magoemon kept a comprehensive record of the eighty-four varieties of cherry tree that he and his family were given or 'collected', including their names and original locations. In this way, not only did he save many cherry varieties from extinction, but his records also offered a rare insight into the state of the *sakura* near the end of the Edo period.

Magoemon's activities in the 1860s and 1870s were to have an even greater impact a few years later. In July 1885 a severe storm struck Tokyo and its surroundings, causing the Arakawa River to flood the village that is today known as Adachi. When extensive repairs of the levees proved necessary, the villagers requested that cherry trees be planted along the riverbanks to stabilise the soil and create shade in the summer.

The village mayor, Kengo Shimizu, who was passionate about preserving Japanese traditions, knew about Magoemon's efforts to preserve cultivated cherries and asked his friend for help. With donations from the villagers, Shimizu bought seventy-eight varieties of cherry – 3,225 saplings in all – from the nursery. Villagers planted them in spring 1886 along about five miles of the Arakawa, and Shimizu asked a budding cherry specialist, Seisaku Funatsu, to preserve and manage these cherries.

By the turn of the century the saplings were mature trees and the Arakawa River started to attract crowds of sightseers, because it was the only location in Japan where the general public could view so many different cultivated varieties of cherry at one time. These Arakawa blossoms perpetuated the diversity of the cherry trees during an era in which the dominance of the *Somei-yoshino* cultivar clouded people's perceptions about the trees' identity. Shimizu's decision to plant diverse cherries was a deliberate act to stem the *Somei-yoshino* tide. Without his enthusiasm and enterprise, Ingram said, many of the finest varieties would have been lost far earlier.

The day after meeting Funatsu at the Arakawa River, on 21 April 1926, Ingram switched his attention to the wild *Yama-zakura* cherries that grew at Koganei in western Tokyo. Many people considered these to be the most varied group of *Yama-zakura* in a single location in all Japan.

There, the sparkling sunshine and cloud-free view of Mount Fuji raised Ingram's spirits as he strolled from a newly opened railway station along a double row of cherry trees. The station

Koganei Avenue, 1926, photographed by Ingram

had been built specifically to meet visitor demand to see Koganei's trees.

Koganei had been established in 1737 by the eighth Tokugawa shōgun, Yoshimune Tokugawa, who had ordered thousands of *Yama-zakura* to be planted along more than three miles of the Tamagawa Jōsui canal. Over time, the area had become famous for its 40- to 50-foot-tall, broad-crowned wild cherry trees bearing single white flowers.

'The Koganei cherries are the best I have seen – many superb specimens,' Ingram wrote. He was particularly struck by the individual characteristics of these wild cherries, which had coppery-red to dull-bronze leaves and mostly white petals, occasionally tinged with pink. Each individual wild cherry was different, with its own petal shape, size and time of blossoming. Like humans, the cherry offspring didn't look exactly the same as their parents, and siblings were dissimilar. This was in sharp contrast to the prolific and identical *Somei-yoshino*.

Ingram was impressed by what was probably the oldest wild cherry tree in Koganei, which had been given its own name, the *Fujimi*. Eleven years earlier the tree's upright trunk had fallen to the ground after a severe storm, and it was thought to be dead. But a new trunk had appeared out of the tree stump, and the *Fujimi* was flowering once again. Desperate to save the tree, Ingram asked Hayashi, the Imperial Hotel's former manager, to send a scion to The Grange.

Everywhere he went, he saw many neglected or dying trees, despite the best efforts of a small group of cherry experts and enthusiasts to save them. Ever the collector, Ingram discovered eight interesting new varieties that he did not possess at the nearby Hanaoka Kōen nursery. He asked its wealthy owner, Teikichi Isomura, a leading member of the Cherry Association, to send him scions. Even if the cherries died in Japan, he was confident that they would be safe in his Kent garden. And now, having visited the cherry havens in Kyoto and Tokyo, it was time for Ingram to discover how the cherries were surviving in the wild.

19. Wild-Cherry Hunting

High in Japan's mountains, away from industrial pollution and human interference, most of the original wild-cherry species of Japan appeared to be thriving. Better still, when Ingram arrived at Hakone, a city two hours west of Tokyo by train, on 26 April 1926, the wild Fuji cherry was in full bloom. As Ingram strode up the mountains above Lake Ashi, to the west of Hakone, he noticed that the vivid red of the outermost petals blended with the white of the inner petals to produce a soft pink effect. About 1,800 feet above the lake, he could see in the distance the reddish leaves and white flowers of the most common wild species, *Yama-zakura*, which created a similar satisfying pink result. On a slope leading to the town of Odawara, about five miles north-east of Hakone, Ingram spied 'fine, upstanding' *Yama-zakura*.

His pilgrimage continued on horseback several days later. An experienced rider since his youth in Westgate-on-Sea, Ingram meandered slowly along 18 miles of narrow, winding forest paths from Lake Shōji, one of five lakes near Mount Fuji, to Fujikawa, a small town to the north-east.

Each turn on the precipitous slopes brought unexpected thrills: a glorious mauve azalea; a towering pile of blossoms on a Sargent cherry tree; an old, time-twisted pine; a snowy crab-apple; a lilac-flowered wisteria sprawling down a rocky cliff. And yet more Sargent cherries. These trees, he wrote, 'were growing on a distant slope, beyond the shadowed depths of an intervening valley – glorious symbols of spring in the midst of a wintry landscape. To complete the picture, the mysterious and mighty cone of Mount Fuji, with its inevitable diadem of clouds, towered to an incredible height in the background, dominating the whole scene.'

In his diary and in later narratives, Ingram wrote about that day – Thursday, 6 May 1926 – with awe and deep emotion:

Viewed against the dazzling blue of the sky, the trees'
flower-laden boughs seemed as though illumined by
a soft glowing light – rosy-hued like the tint of dawn-
flushed snow. Even the trunks and branches were things
to admire. Naturally glossy, where a shaft of sunlight fell
upon their surface, they were transformed into glistening
columns of burnished bronze.

Here and there, in the pit of the hollows, the eye could
make out a nestling, thatched-roofed village, set about
with green terraced fields. But the huddled, fantastically
steep mountains claimed the landscape, and man was
here a mere incident. For long I sat and let the beauty of
the scene soak into my soul.

This transcendental moment was heightened by the melding of Ingram's two great passions – birds and blossoms:

> As I stood there, spellbound, I heard a sudden, almost startling, outburst of melody from a nearby grove of bamboo. Somewhere out of the bosky gorge rose the muffled notes of a dove and then the loud, long-drawn liquid pipe of the *uguisu*, the little russet-coloured nightingale of Japan, ending with a strange, sudden flourish. It would be difficult to conceive a more fitting association of sight and sound in the heart of that vast and lonely forest.

Weeks into his trip, when Collingwood Ingram arrived in Nikko, 95 miles due north of Tokyo and 4,000 feet above sea level, thick snow enveloped the city. The woodlands here were as stark and bare as in midwinter. In September 1902, on his first visit to Japan, Ingram had been entranced by this city's Buddhist temples and by the mausoleum of Ieyasu Tokugawa and his grandson, Iemitsu, the first and third shōguns of the Tokugawa shogunate. This time, after a brief visit to the temples, he hiked through the leafless woods in search of blooming cherries. His hiking partner was Ishiguro Suzuki from the Yokohama Nursery, a plump man with a fixed smile who was wearing a European city suit.

The men walked for miles along mountain tracks and beside a swollen river, before entering a dense deciduous forest. On a mountain ridge at about 2,000 feet, Ingram was suddenly spellbound. In front of him was a superlative specimen of a Sargent cherry 'in its full vernal glory, with every bough and every twig wreathed in soft shell-pink blossom'. When Ingram dashed off to collect some tiny seedlings, he lost Suzuki. But he gained a superb cherry, which grew for years in a corner of his croquet lawn at The Grange.

On a separate walk from Nikko to the picturesque Lake Chūzenji, 15 miles west of the city, Ingram collected more Sargent seedlings to take back to England and found an ancient specimen with unusually large flowers at about 2,500 feet above sea level. Next on the list was a visit to an avenue of healthy, evenly matched and wide-spreading *Somei-yoshino* cherries in Utsunomiya, east of Nikko. Despite their ubiquity, Ingram still appreciated the 'floriferous and beautiful' trees, in the same way that a dog-lover would still love a Labrador even if the majority of dogs in the world came from that one breed.

From Utsunomiya, Ingram travelled more than 200 miles north by car to the pine-clad islets of Matsushima Bay on Japan's east coast, where he hired a motor boat and drifted out to sea for a few miles. There, looking back at Matsushima village through his binoculars, he spotted a solid mass of rosy-pink on one side of a whitish cloud of *Somei-yoshino* cherries. Excited by the unusual sight, Ingram rushed back to shore, to discover one of the most magnificent and largest *Edo-higan* wild cherry trees he had ever seen. Planted on an incline next to a small deserted shrine, the tree was at least 40 feet high with an 18-foot girth at its base. 'Every twig was closely packed with blossom — the individual flowers being of relatively good shape and size,' he wrote. 'I have rarely seen a more beautiful object.'

Out in the countryside, away from the cities, Ingram was on a roll. In the straggling village of Kami Yoshida, for instance, at the foot of Mount Fuji, he made one of his most exciting discoveries. There, in the garden of a house near the Osakabe Hotel, towering above a tall wooden fence, stood a tree with narrow leaves and bunched clusters of double mauve-pink blossoms with close to 100 petals. It resembled a cherry that E.H. Wilson had described in 1913, but was clearly a different variety. Nor was it included in Professor Miyoshi's classification guide. Ingram's immediate reaction was to work out how to spirit cuttings of the tree to England.

Fate was on his side. Nineteen years earlier, on his honeymoon, he had visited this very village while hunting birds, and he remembered meeting there a one-legged war hero whose parents ran the Osakabe Hotel. That man, who had lost a limb during the Russo-Japanese War, was still alive, a villager told Ingram. Indeed he was now running the hotel. And his hobby was gardening! In typical Ingram fashion, he convinced the innkeeper to send him scions from the tree, in exchange for one yen to cover the postage. By 1929 a couple of sturdy offspring were growing in Benenden.

Ingram dubbed the plant *Asano*, after the hero of the *Forty-Seven Rōnin* saga. The story was that in April 1701 a *daimyō* from western Japan, Naganori Asano, had been ordered by the fifth Tokugawa shōgun, Tsunayoshi Tokugawa, to commit ritual suicide, or *hara-kiri*, after wounding an envoy from the imperial family in Edo Castle because he thought he had been insulted. In revenge, forty-seven of Asano's samurai retainers, now leaderless and known as *rōnin*, decided to kill the envoy themselves. They were caught and also slit their bellies. In the *kabuki* play about the incident, *Chūshingura*,

Asano's final act was to read the death poem that he wrote before committing *hara-kiri*:

> More than the cherry blossoms
> Inviting a wind to blow them away
> I am wondering what to do
> With the remaining springtime.

Although Ingram never said why he called this particular cherry *Asano*, it probably reflected his interest in samurai stories. Today the *Asano* cherry is a popular variety in Britain and forms the centrepiece of the Asano Avenue at Kew Gardens in London.

20. Saving the *Sakura*

Collingwood Ingram hated giving public speeches. Additionally he believed it was impertinent to share his opinions about cherry blossoms with the Japanese, given their more than 2,000-year heritage and his less than seven-year apprenticeship as an English '*sakuramori*' or 'cherry guardian' — the Japanese name given to collectors and protectors of the trees. So throughout his 1926 trip he had kept most of his views to his diary. There they might have stayed, had not Aisaku Hayashi, the founder of the Cherry Association, asked Ingram to speak frankly about the state of the nation's cherries to an audience of royalty, business leaders and top bureaucrats.

There was a lot at stake. In April 1917 Hayashi had formed the Cherry Association, *Sakura No Kai*, as a way to 'make known to the world that the cherry blossom is the national flower of Japan, to make the flowers worthy of our pride … and to serve our country by protecting our cherry trees'. At the time Hayashi was general manager of the Imperial Hotel, and cherries were high on the list of things that wealthy Westerners who stayed at his hotel wanted to see in Japan.

Unfortunately Hayashi knew little about cherry trees, and even less about the deterioration and loss of some varieties. Undeterred, he went straight to the top, to Eiichi Shibusawa, an industrialist known as the 'father of Japanese capitalism' who had been involved in virtually all facets of Japan's meteoric rise. Not only had he been the head of the Finance Ministry, president of a major bank and founder of the Tokyo Stock Exchange in 1878, but he also had helped to form companies making and selling everything from cement to paper. What is more, Shibusawa had been chairman of a company set up to build the Imperial Hotel.

Opened in 1890, the 57-room hotel overlooked the Imperial Palace's cherry-bedecked outer moat and immediately became Tokyo's finest Western-style hotel. Each room even had its own telephone — a rarity at the time. Yet despite the luxury, business was bad. The rooms were expensive and there were few foreign visitors to Japan. Moreover, the Swiss manager's hiring of a French chef and other foreigners was not paying off. So in 1909 Shibusawa decided to fire all the non-Japanese staff and to head-hunt a Japanese general manager. He found Hayashi, a 36-year-old fluent English

The Cherry Association's Cherry Dance Party, 1919

speaker who was managing a Japanese arts and antique store in New York.

Over the following ten years Hayashi transformed the Imperial Hotel, and it didn't take much for him to convince Shibusawa to finance the Cherry Association and provide it with office space. In turn, Shibusawa persuaded close friends to back the association. The honorary president was Marquis Yorimichi Tokugawa, a Cambridge-educated nobleman descended from the Tokugawa shōgun's family. Within weeks, Tokyo's wealthiest and most well-connected citizens were members of the group, under the chairmanship of Shibusawa. Other directors included Professor Miyoshi, Seisaku Funatsu, Hayashi and Kiyoshi Inoshita, a cherry expert in Tokyo's parks department.

At the first gathering, on 23 April 1917, about 200 new members met in the Imperial Hotel, which was then being redesigned by Hayashi's friend, the American architect Frank Lloyd Wright, who would also become a member. To highlight the cherries' diversity, the group exhibited forty-three different varieties and agreed to launch an annual journal, called, predictably, *Sakura*.

Yet while Aisaku Hayashi was always an active member of the association, his tenure with the Imperial Hotel was cut short. Massive cost and schedule overruns during construction of Wright's new building damaged his relationships with the architect and with the board. The final straw came in April 1922 when fire swept through the old building, killing a Greek resident who had run back into the hotel to save his cat. The hotel's board, including Hayashi, resigned, to take responsibility for the fire. Frank Lloyd Wright left Japan three months later.

The Cherry Association's directors, particularly Funatsu, Miyoshi and Hayashi, had a genuine concern about the cherries' future and repeatedly lamented both their decline and the Japanese people's lack of interest in them. During *hanami*, the cherry-viewing season, the Japanese 'maintain a pretext of flower-hunting and sightseeing, but their real interest is in satisfying their inner selves with food and drink,' they wrote in the first edition of the *Sakura* journal. And they continued:

> The cherry blossom symbolises the outburst of our
> national spirit. It is a symbol representing the whole
> Empire. How shameful it is that we must say that the

present tendency of the mass, if left undirected, is to disregard this beautiful flower in the Land of the Gods.

It has come about then that many famous places for the cherry blossoms are in a stage of decay and the new trees planted in place of the old ones are set out without order or design worthy of our traditions. If this neglect continues we fear that the famous beauty of the blossoms of the cherry land of the Orient shall be gradually ruined.

In the second *Sakura* journal, published in April 1919, Hayashi was even more pessimistic. 'We are face to face with the facts of the decay of the trees everywhere,' he wrote, 'and the extinction of this crowning glory of Japan is in sight.' He mentioned that about 500 of the trees at the famous Arakawa riverbank had been cut down and burned into charcoal, probably because of construction after the river flooded. Other trees had been vandalised. And he continued:

No easy task is before us. And yet we will go on, nothing daunted. There is so much to do and so little done. Mountains, hills, country roads and river banks all over the Empire are anxiously waiting our beautifying and magic touch. Our souls, too, need reawakening. Some day that must come. What a reawakening that will be!

E.H. Wilson, the plant-hunter, had noted a year earlier, after a trip to Japan in 1918, that the trees' health had deteriorated in many places, especially in Tokyo. Not only did the trees need spraying with insecticide to kill pests, Wilson wrote, but some also needed protecting from vandals, and young and healthy trees needed to be planted to replace old and decrepit specimens. 'The cherry blossom is essentially Japanese and its passing would be a national calamity – nothing less,' he wrote.

However, for many of the elite who attended the Cherry Association's meetings, the society's increasingly strident calls to save the cherries fell on deaf ears. The annual get-togethers were less about the cherries than about making and maintaining contacts. All believed that the cherry should be preserved and venerated as a national flower. But it is less clear whether most of the crème de la crème had a genuine interest in protecting the different varieties, or whether they even realised that a problem existed. After all, in parks throughout the land millions of people gathered each spring to enjoy *hanami* beneath the pink-and-white blossoms of the *Somei-yoshino* cultivar. On the surface, at least, the trees seemed to be thriving.

21. Ingram's Warning

For the meeting at which Ingram had been invited to speak, about 150 of Japan's most distinguished titans, plus a bevy of reporters and photographers, gathered on Tuesday, 27 April 1926 at *Kokumin Shimbun*, an influential pro-government newspaper and the forerunner of today's *Tokyo* newspaper.

Introducing Ingram to the association's members, Professor Miyoshi said that he was encouraged by Ingram's visit to Japan to gather cherry varieties and preserve them in his English garden. But he was also distressed about their condition in their homeland. 'Many cherry trees in Japan are in a deplorable condition,' Miyoshi said. 'The trees that were sent to the Potomac are well looked after. And as a result of their preservation, they are growing better in Washington than those in Japan. It's an ironic reverse situation from what you might have expected.'

When Ingram began to speak, with Hayashi translating, he continued with that theme. Glancing at the notes he had written the evening before on the Imperial Hotel's creamy stationery, Ingram asked a blunt question and then proceeded to answer it. 'Why is it,' he said, 'that your flowering cherries often seem to do better in England than in their native country? I would like to confess to a feeling of disappointment with regard to the size and condition of some of the cherry trees growing in your parks and other public places. I do not suppose that the alien soil and climate [in England] is more congenial. We have to seek other causes.'

The audience listened politely as he listed the reasons. It was presumptuous, of course, for an English cherry-tree neophyte to question Japanese cherry-growing techniques that had evolved over hundreds of years. Yet Ingram wanted to address what he thought were tangible problems. 'My knowledge of your methods of cultivation is, of course, of a very superficial nature,' he said. 'I offer them for what they are worth — you can accept or reject them as you please.'

First, Ingram said, the flowering cherries didn't live long because of the common Japanese practice of using a cherry variety called *Mazakura* as a stock plant to propagate varieties. *Mazakura* was a 'feeble and short-lived variety,' he said, and he encouraged the audience to use a more vigorous form of wild cherry. In England, he grafted most of his cherries onto *Prunus avium*, the wild English cherry tree, which helped the new cultivar adapt more easily to Britain, because it was used to the soil and the climate.

Second, he continued, Japanese gardeners were planting mature trees, rather than saplings. This was a 'grave error,' he said, because older trees were more susceptible to disease and insect attack. Another cause of the cherries' ill-health derived from planting trees on the same site for 'hundreds, perhaps thousands, of years'. The soil became 'cherry-sick,' he said, and was unable to nourish new plants.

After listing these problems, he delivered a stark warning:

'Long before aesthetic Japan became contaminated by the hustle and bustle of the Western races – your people produced, no doubt by careful and painstaking selection, an amazing number of varieties. In recent years, not only has there been no attempt whatever made to improve these varieties, but many of them are in serious danger of extinction.

'Were it not for a handful of enthusiasts – the moving spirits of the *Sakura No Kai* – it is scarcely an exaggeration to say that in another fifty years you would have permanently lost most of the varieties that had been evolved with such loving care by your ancestors.

'I feel confident that in years to come, the Japanese will have to seek some of their best sorts in Europe and America.'

The hall was silent as Ingram concluded. He told his audience of two beautiful cherry varieties growing in his English garden, which he had not been able to find in Japan. He promised to return them to their native home: 'If I am able to restore these to your country, it will indeed be a proud moment for me, and I will feel that I have repaid with small mite the wealth of beauty my garden has derived from your lovely country.'

The irony was that on the very day of Ingram's speech, the two rare Japanese varieties that he had mentioned were in bloom in his Kent garden. Along one pathway at The Grange stood some stately *Taihaku* trees, their branches laden with large, single white flowers. Nearby some *Daikoku* cherry trees bore handsome double-pink blossoms. Both were Japanese varieties that had been exported to Europe, but by 1926 they seemed to have disappeared completely from their native land.

From that moment on, reintroducing these two cherries to Japan and saving other dying varieties became Ingram's priorities. Quite how he would accomplish these goals, he had yet to figure out.

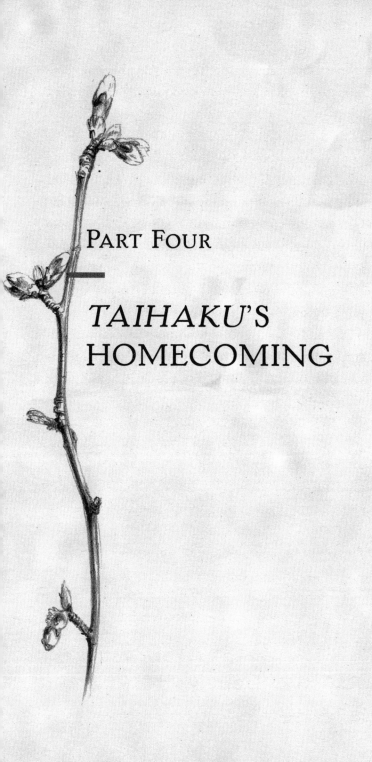

PART FOUR

TAIHAKU'S HOMECOMING

22. The Restoration Quest

Collingwood Ingram left his first-class cabin aboard the *Empress of Scotland* on Thursday, 1 July 1926, and tested his land legs on the docks at Southampton. The journey back to England from Japan – by ship from Nagasaki to Vancouver, then a transcontinental railway across Canada and from there another ship from Quebec – had taken forty days.

On the long trip home, Ingram reflected on his cherry pilgrimage to Japan. Most wild cherries in the mountains were in good shape, largely because they were in remote locations. Yet their man-made offspring, aside from the ubiquitous *Somei-yoshino*, were endangered. Worse, only a handful of cherry-lovers realised this, and even fewer were looking for a remedy. And that included the business and government leaders whom he had addressed in his Tokyo speech. In fairness, he wrote later, their muted reaction was in part perhaps because many attendees didn't understand English. 'This was a pity, for a few home truths might have had a salutary effect and stimulated an interest in the rarer – and disappearing – varieties.'

In fact Ingram's speech had been widely reported. The Japanese enjoyed reading what visiting foreign 'experts' said about their country, even when the comments were critical. The headline in the *Tokyo Shimbun* newspaper said it all: 'Japanese Are Indifferent Towards Preservation of Cherries'. By today's standards, Ingram's speech was balanced and moderate. However, in 1926 his reprimand for what he regarded as Japan's avoidable failure to protect its national flower was seen as a harsh condemnation.

Yet his speech had little, if any, impact. In 1926 the government and business priorities were to rebuild after the massive 1923 earthquake, revive the battered economy and maintain stability. Politically, Japan was becoming increasingly intolerant. A year earlier, for instance, the government had passed the Maintenance of the Public Order Act, to make it easier to monitor and outlaw socialists and communists, and to expand the powers of the *Tokkō* police force, also known as the Thought Police. In that atmosphere, reviving the cherries was an afterthought. There was no money, no urgency and no political will.

Given the circumstances, Ingram decided it was up to him to make a difference. In 1919, when he had first become interested in cherry trees, his goals had been to become an acknowledged *sakura* expert and to collect as many varieties as possible. Now, well on his way to achieving those aims, he added three more ambitions: to return extinct cherry varieties to Japan from England, to propagate Japanese varieties in his Benenden garden and to spread the cherry tree throughout the UK and beyond.

Yet during the time Ingram had been out of the country, the British political landscape had also changed. In May 1926 a frustrated Trades Union Congress had ordered a General Strike of 1.7 million workers, after the owners of coal mines had tried to get miners to work longer hours for less pay. Prime Minister Stanley Baldwin, leader of the Conservative Party, proclaimed a state of emergency and refused to negotiate with the unions until the strike was called off.

The stand-off highlighted the gulf between owners and workers as much as it did the gulf between the Conservative and Labour Parties for which they voted. Ingram was a Conservative — a country-dwelling, nature-loving semi-aristocrat from a privileged nouveau-riche background. He couldn't relate to the hard-scrabble life of the working classes from which his grandfather, Herbert, had escaped after founding the *Illustrated London News*. Yet neither did he feel comfortable with the politicians, industrialists and businessmen who valued acquisitions over aesthetics.

In the turmoil Ingram didn't so much retreat from British society as tread his own path. Back at The Grange with Florence, his uncomplaining wife, he resumed his horticultural routine. The couple's four children, now aged between nine and eighteen years, spent their formative years at private boarding schools. They needed little supervision — not that Ingram had ever been much of a hands-on father. While Ingram enjoyed his gourmet food and wine, he had no interest in dressing up or in social niceties. So after returning from Japan, he turned all his attention to the one constant in his life that decade: the cherries.

By 6 a.m. each day he would be out in his garden, his jacket pockets weighed down by a steel-backed trowel, razor-sharp knife and fine-tipped embroidery scissors. Around his waist, a thin leather belt held up the same pair of trousers that he wore until they were threadbare. Hat perched atop his long, unkempt greying hair, he would check on his flowering cherries one by one, rain or shine.

What *was* their attraction? Above all, Ingram said, it was their 'refined charm, when in bloom, and a delicacy of colour and form that appeal to one's aesthetic sense in a way that others can never do'. Many other flowers were 'so gross and gaudy as to look no longer like things of God's creation'.

During the winter of 1926, six months after his return from Japan, the cherry tree scions that he had asked for started to arrive at The Grange. The cuttings came from all over Japan: from the Yokohama Nursery Company, from Tsuneo Kajūji in Kyoto, from Isomura's horticultural shop in Koganei, and from the one-legged soldier whom he had met in Kami Yoshida. Opening each crate was always an occasion, because he could never be sure what it would contain.

Everyone had kept their promises to send scions. And each cherry enthusiast had devised a way to keep them alive during the five- to six-week sea journey to England, often by attaching water-filled moss at the base of each scion.

One particular crate, which arrived in March 1927 from Seisaku Funatsu, contained thirty cherry-tree scions from seven different varieties growing on the Arakawa River: *Ichiyō, Komatsunagi, Bendono, Shirayuki, Fukurokuju, Temari* and *Shōgetsu*. A further sixty-six scions came from Aisaku Hayashi, the former manager of the Imperial Hotel in Tokyo, in February 1928. His crate contained scions of five types of *Yama-zakura* from the Koganei avenue in western Tokyo, one of which was the ancient *Fujimi-zakura* tree that Ingram had so admired in Koganei. The Koganei cuttings had been selected and collected by Yōichi Aikawa, the engineer from the parks department in Tokyo who had shown Ingram around Koganei in 1926 with Hayashi. As Hayashi wrote to Ingram, the scions had travelled in style, aboard the luxury Japanese passenger ship, the *Mishima Maru*, to Vancouver.

Not all the plants survived. In Tsutsujigaoka Park on the outskirts of Sendai, a city 225 miles north-east of Tokyo, Ingram had been fascinated by the deep-pink flowers of a weeping cherry tree, *Yae-beni-shidare*. He called it the most beautiful of its type that he had ever seen. But when scions arrived that winter at The Grange, he was heartbroken. After carefully unwrapping the packaging, he saw immediately that the thin branches inside had perished. The *Fujimi-zakura* scions from Koganei apparently suffered a similar fate.

Nonetheless, most of the plants sent from Japan thrived in his garden, and eventually throughout the world. To keep the cherries alive, Ingram usually placed them in moistened moss inside a shallow lidless box. Once they had sprouted, he would sprinkle the new foliage using an old hairbrush twice a day, dipping the bristles in water and then shaking them over the plants to imitate anything from a drizzle to a thunderstorm. It was a technique he had learned from his friend Edward Augustus Bowles, author of a gardening trilogy and owner of Myddelton House in Enfield, Middlesex.

To propagate the plants, Ingram would graft the variety onto the native English *Prunus avium* species. By trial and error over several years, he developed methods to increase the success rate of these grafts. Sometimes he made a T-shaped incision in the trunk of the *Prunus avium* and then cut a notch in the vertical section, into which to ram the short stem of the scion, binding the connection with insulating tape. At other times he made a slanting cut on the upper side of a branch and thrust the wedge-shaped end of a scion into the cleft. His methods required knowledge of plant physiology, nimble fingers and a degree of dexterity that came from years of practice.

Ingram sent seeds and scions from these trees to countless correspondents without charge. From 1928 to 1931, for example, he shipped scions of at least twenty cherry varieties to the US

Department of Agriculture in Washington D.C. He had hoped that these could be distributed further around the United States, but that was impossible because of plant quarantine regulations.

For friends outside England, Ingram would often coat the entire surface of the scion with a film of paraffin wax to seal the pores of the wood, preventing it from withering. His sole motive was to save the cherries by distributing them for all to plant and admire. By the early 1930s The Grange had become a cherry-blossom trading post, with Ingram at the helm, receiving and dispersing seeds and scions to and from both Japan and the United States, to an ever-increasing number of horticulturalists. Yet in spite of these altruistic achievements, Ingram gained most satisfaction from another accomplishment.

This was the reintroduction in the early 1930s to Japan of the *Taihaku* cherry. Returning this living plant to the land of its birth was a difficult and frustrating process. It took several years and stirred strong emotions in Japan. How could it be that one of the country's most magnificent cherry blossoms was saved from possible extinction by a foreigner?

23. *Taihaku*'s Homecoming

One day in 1923, three years after Ingram had begun his quest to collect as many different cherry-tree varieties as he could find, he visited the Greyfriars Estate in Winchelsea, a coastal town in East Sussex 15 miles south-east of Benenden. There a fellow cherry enthusiast called Annie Freeman lived with her barrister husband, George Mallows Freeman, and their three children. In 1899, when Annie was visiting Provence, she had met a Frenchman who described some outstandingly beautiful cherries that he had seen in Japan. Annie, who had already imported several rare flowering cherries from Asia, contacted a friend of this Frenchman in Japan. Through him, she imported another small collection of cherries that matched the Frenchman's description of these trees.

Among the plants that Annie Freeman had bought were four cherries, from which Ingram took scions back to The Grange. One was a variety with large pink, almost purplish, double-blossom flowers that Ingram later named *Daikoku*. Two others were *Tora-no-o* − a cherry with single light-pink petals, also known as the Tiger's Tail cherry − and another single-petalled variety called *Chōshū-hizakura*.

Yet what most attracted Ingram was a fourth cherry with exceptionally large snow-white blossoms, whose name was unknown. Growing in Annie's shrubbery, the tree was in a sad state, although a couple of its boughs were still alive. The blossoms, Ingram reported, were huge: up to 2.4 inches in diameter, with leaves as long as 7.5 inches. Ingram had never seen this cherry before, but he immediately recognised 'the rarity and remarkable beauty of this variety' and took home a few twigs for grafting.

The results were spectacular. And when Ingram's friend, Duke Nobusuke Takatsukasa, visited The Grange in the spring of 1925 he dubbed the tree *Taihaku* − the 'Great White Cherry'. The name − 'tai' (great) and 'haku' (white) − summed up the size and colour of the blossom, although it didn't quite capture the beauty or sense of tranquillity that had attracted Ingram.

A year later, on the afternoon of Tuesday, 20 April 1926, Ingram was sitting in the living room of the cherry expert

Seisaku Funatsu, near the Arakawa River in Tokyo. Eating *wagashi*, traditional Japanese sweets, and drinking green tea with his kimono-clad host, he had carefully studied Funatsu's prized album of local cherry blossoms, called *Kōhoku Ōfu* (Cherry Trees of the *Kōhoku* Region). It consisted of fifty-seven water-colours of cherry trees on high-quality *washi* (Japanese paper). Funatsu had commissioned the paintings from an artist called Kōkichi Tsunoi, over a period of seven years from 1913, so that there would be a record of the cherry trees planted along the riverbanks.

One by one, Funatsu explained the characteristics of each variety as Ingram, himself an adept painter, marvelled at the minute details in each sketch. Reaching the end of the album, Funatsu left the room and returned carrying a long and narrow *kakemono*, or scroll, which had been hanging elsewhere in his house. He unrolled the scroll carefully. It bore a vivid paint-ing of a cherry plant with immense white blossoms. Funatsu said that the plant was a variety called *Akatsuki*, a relative of Ōshima cherry. But he added wistfully, 'This is the cherry that my great-grandfather painted more than 130 years ago. We used to see it near Kyoto but it seems to be extinct. I can't find it anywhere any more.'

Ingram gasped. From the size and colour of the flowers to its young copper-red leaves, the blossom on the scroll was undoubtedly *Taihaku*. 'This cherry is growing in my garden in Kent!' Ingram blurted out. It seemed like fate. Funatsu 'was clearly incredulous, but his good manners forbade any open expression of doubt,' Ingram wrote.

Funatsu said nothing. He just smiled and bowed deeply and courteously towards the Englishman. 'Piqued by his obvious doubt, I there and then resolved to convince him,' Ingram wrote. He swore to himself that he would return the blossom to Japan, even if Funatsu didn't believe his story. A week later, in his speech to the Cherry Association in Tokyo, Ingram repeated his vow to reintroduce *Taihaku*. He also said he would return *Daikoku*, which he had named after a Buddhist god famed for bringing good fortune and happiness. At the time an image of the portly deity featured on Japan's one-yen note. Like *Taihaku*, this variety no longer existed in Japan.

Ingram focused most of his attention on returning *Taihaku*, his favourite specimen and 'by far the most beautiful of all the white cherries'. Since it was only possible to obtain scions in winter, he had to wait months before putting some cuttings on a ship. So in early 1927 Ingram sent scions of *Taihaku* and *Daikoku* to the Cherry Association in Tokyo. Regrettably, the society reported, both varieties 'withered away'. From this point on, *Daikoku*'s fate in Japan is unclear. But the following year, 1928, Ingram sent several *Taihaku* scions to a 40-year-old cherry enthusiast called Masuhiko Kayama, the author of *Kyoto no Sakura* ('Cherry Blossoms of Kyoto'). Kayama and Ingram had been introduced through the correspondence of Count Tsuneo Kajūji. The two never actually met.

A gregarious botany teacher with a loud, distinctive voice, Kayama was the deputy headmaster of a Kyoto girls' school. He was descended from a family of *tera-zamurai,* a class of samurai responsible for protecting and maintaining temples during the Edo period. The Kayama family protected one of the nation's most venerated sanctuaries, the Ninna-ji Temple in north-west Kyoto, now a World Heritage Site. Founded in AD 888, it was the head temple of the Omuro school of Buddhism's Shingon sect. It also had extremely close links to the imperial family, who often spent summers there. Additionally, for more than a thousand years the Japanese emperors had frequently sent a son to the temple to serve as the head priest.

As well as a magnificent five-storey pagoda, Ninna-ji was known for its cherry trees, so Ingram had little doubt that *Taihaku* would fare well in the temple grounds, if it was grafted successfully. Kayama asked a friend called Tōemon Sano for help in grafting Ingram's scions. Sano's family owned a nursery and had been in charge of gardening at the Ninna-ji Temple for more than 500 years. His name, Tōemon Sano, had been passed down from one head of the family to the next for sixteen generations. It's not uncommon in Japan for one family to be associated for centuries with a specialist role, be it making pottery, swords, kimonos or flower-arranging and the tea ceremony. Usually the eldest son inherits his father's title as master of the discipline.

The last three generations of Sano's family had had a special interest in cherry trees and had become known as *sakuramori* or cherry guardians. They were the fourteenth successive Tōemon Sano, born in 1874; his eldest son, the fifteenth Tōemon Sano, who lived from 1900 to 1981; and the sixteenth Tōemon Sano, born in 1928.

Both the fourteenth and fifteenth Tōemon Sano collected cherry trees throughout Japan. Besides protecting the trees at Ninna-ji Temple, the fifteenth Tōemon Sano was also revered for growing one of Kyoto's most famous trees, an iconic weeping cherry or *Shidare-zakura*, in Maruyama Park.

The sixteenth Tōemon Sano, the current *sakuramori*, continued the story when I visited his home at Uetoh Zohen in Kyoto in December 2014. Uetoh Zohen is the name of the family's

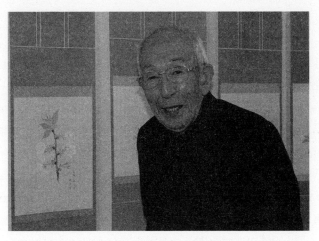

The sixteenth Tōemon Sano

landscape-gardening company, and all sixteen generations of the family have lived in the residence. Sano was eighty-six years old when I visited, but the decades fell from his suntanned face as he recounted the *Taihaku* tale, which had been passed on from his grandfather and father.

'Masuhiko Kayama and Grandpa were very close friends,' the sixteenth Sano said in his soft Kyoto accent as we drank green tea in a parlour, surrounded by framed drawings of cherry blossoms that his forefathers had collected. 'Grandpa decided to graft the British *Taihaku* because Kayama-san asked him to. But the scions were dead on arrival for three years in a row, from 1928 to 1930, and neither Kayama-san nor Grandpa could figure out why.'

According to Sano, that was when his father, who was in his early thirties at the time, stepped in. He thought that conditions on board the ship must be too dry, so the *Taihaku* cuttings were not getting enough water during the voyage to Japan. The fifteenth Sano suggested that Ingram send the cuttings embedded in daikon radishes, for moisture. The cuttings duly arrived, inserted in the vegetables, in 1931.

'Kayama-san, Grandpa and Dad were so excited when the cuttings arrived. But when they unpacked them, they saw they were rotten. There were long shoots sprouting from them on all sides. It was so disappointing. This time, they blamed the problem on too much water, not too little.'

'What happened next?' I asked.

Sano sat up straight in his chair and recounted the conversation as though it had just occurred. '"I've got it. The problem is crossing the Equator!" Dad shouted. "The heat near the Equator must have made the cuttings grow shoots. Then, when they arrived in Japan, the cold must have killed them."'

At the time, most scions sent from Japan to Britain travelled on a northern route, usually via Vancouver or San Francisco. This was the route home that Ingram had taken in 1902 and 1926. However, the *Taihaku* cuttings had travelled on a southern route, as did most passengers between Britain and Asia in the early twentieth century. They usually went on ships via the Suez Canal and the Indian Ocean.

To avoid the intense heat of the southern route, Sano's grandpa suggested that Ingram try sending the scions on the Trans-Siberian Railway instead. So in early 1932 Ingram took some small potatoes, cut them in half and pressed the bottom end of the *Taihaku* branch into the exposed surface. He thought the moderate levels of moisture and nutrients in the potatoes would be an effective way of transporting the scions, and arranged for them to enjoy a leisurely trip across Siberia.

After crossing the Siberian tundra, where the outside temperature was below freezing, the cuttings were taken from

Vladivostok's railway terminal to a harbour and then crossed the Sea of Japan by ship. After an overnight train to Kyoto, they arrived at the Sano residence. It was now almost six years since Ingram had pledged to return *Taihaku*. Once again, Kayama and the Sanos gathered at their family home to unwrap the plant.

The sixteenth Sano suddenly stopped talking and looked up at me, allowing silence to fill the room in which we were sitting. 'So what did they find?' I asked him eagerly. Sano took a sip of tea and then a child-like grin brightened his weather-beaten face as he prepared to deliver the punchline. 'They were alive, Abe-san! Alive! A common potato saved *Taihaku*.'

Sano's father set about grafting the cuttings onto Ōshima cherry stock. Within weeks one graft started to grow. When the sapling was about three years old, he successfully took further scions from the plant and spread *Taihaku* beyond his lush gardens. At last the plant was home. And safe.

One tree was planted in the school yard where Kayama worked, and is still alive, according to his grandson. Another lives in his garden. A third was planted at the Ninna-ji Temple and a fourth at Hirano Shrine, where gardeners erected a sign explaining the cherry's return from Britain.

To Sano, *Taihaku* retained some of its British heritage. 'It's not just that it's a pure white flower,' he told me. 'It's dignified, elegant. Even though it was originally brought to Britain from Japan, it seems to have acquired a gentlemanly grace while it was over there.'

However, not everyone cheered *Taihaku*'s homecoming, for the political and economic mood in Japan had grown dark. The Anglo-Japanese Alliance, a bilateral cornerstone

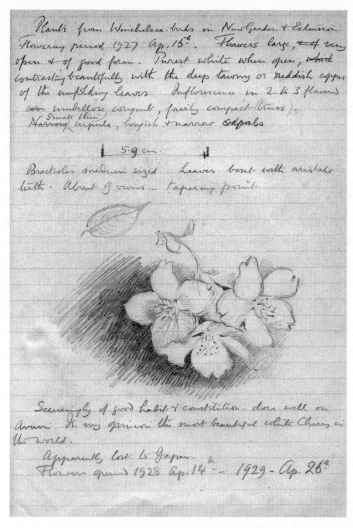

Taihaku: *a page from Ingram's notebook*

for twenty years, officially ended in 1923. As numerous small Japanese banks collapsed because of a financial crisis in 1927 and the subsequent Great Depression, economic power shifted to the four major conglomerates – Mitsui, Mitsubishi, Sumitomo and Yasuda – that dominated business in Japan.

At the same time, political power shifted towards national-ists and the military, and Britain was increasingly seen as a potential enemy.

For some cherry aficionados, it was disgraceful for Japan even to think about receiving a cherry tree from an English-man. In 1928, for instance, after Ingram wrote to Kayama asking if he would like a *Taihaku* scion, Kayama had made an unannounced visit to the Osaka home of Shintarō Sasabe, an independently wealthy cherry-collector known as the 'Cherry Man'. Like Ingram, Sasabe was determined, eccentric and obsessive about cherry blossoms. According to Sasabe's autobiography, Kayama told Sasabe that Ingram had con-tacted him, asking whether he would like to receive a *Taihaku* tree that he said was extinct in Japan. Kayama said he was conflicted.

'I feel it might be a disgrace to accept Ingram's offer because Japan is the home of the cherry blossom. But I also feel it would be a shame not to take it. What do you think?' Kayama asked Sasabe.

Sasabe was blunt. 'It's better not to,' he said. 'How does Ingram know for sure that *Taihaku* is extinct here? He only visited a few places in Japan, places where there were luxury hotels. Although our cherry varieties are disappearing, we'll still find *Taihaku* somewhere. Thank him politely, but don't accept his offer. It's unworthy.'

Sasabe, a cherry guardian who despised the *Somei-yoshino* variety, later claimed that he had found and grafted a cherry that resembled *Taihaku*. The war prevented him from com-paring the varieties to see whether they were identical, he said. He also stated that the sign at Hirano Shrine about

Taihaku was removed during the war because England was the enemy.

Nonetheless, Kayama and the older Sanos ignored Sasabe's advice. It was more important, they believed, to welcome *Taihaku* home than to worry about how it got there. The men were proud of Japan's history of developing different cherry trees and were passionate about continuing that tradition. It was an attitude, Sano told me, that his father and grandfather shared with Ingram.

Wandering around his garden in the cold December air, Sano pointed to a mature leafless tree. It was the offspring of the *Taihaku* that his father had grafted. Sad to say, Seisaku Funatsu, the cherry specialist whose serendipitous scroll had prompted *Taihaku*'s return, had died in 1929 without knowing of the tree's homecoming. This was something Ingram regretted for the rest of his life. The fourteenth Tōemon Sano also died, in 1934, before the tree could show its blossoming beauty.

While this was a bittersweet ending to *Taihaku*'s homecoming, Collingwood Ingram could not have been more proud of the outcome. 'From that tiny nucleus [of *Taihaku* trees at The Grange], tens of thousands of trees have been propagated,' he wrote in 1948. '*Taihaku* has now encircled the world. From a chance meeting in a small Provençal town, the loveliest of all Japanese *Sato-zakura* [cultivated varieties] was miraculously saved from extinction.'

When Funatsu died, Ingram lost a friend who had appreciated cherries with the same passion that he had. But he found another Japanese cherry confidant in Masuhiko Kayama. After *Taihaku* had been successfully grafted in Kyoto, Kayama started writing long letters in English to Ingram, which were often sent with cherry scions taken from the ancient trees at the Ninna-ji Temple and Hirano Shrine.

Among the trove of documents kept by Ingram's grandson-in-law, Ernest Pollard, was a letter from Kayama to Ingram dated 27 July 1934. The letter suggested that the majority of the scions he had sent to Ingram had failed to grow at The Grange or had died en route. But Ingram did manage to graft and propagate at least two of Kayama's varieties in his garden: the large-flowered pinkish-white *Ariake* (Dawn) and *Shōgetsu* (Moonlight on the Pine Trees). The following winter, Kayama wrote, he was going to send more grafts two ways – one via Siberia, using Ingram's potato method, and the other via Canada in a Thermos flask.

An earlier letter from Kayama contained a poem that he had written in 1932 in English, called 'Song to the Cherries', which highlighted the blossoms' close links with Britain. What the poem lacked in literary finesse, it made up for in fervour, at a time when an increasingly militaristic Japan was loosening its ties to its island counterpart in Europe:

> Oh cherries, cherries, my dear cherries
> Go across two oceans and a continent safely.
> You don't know England yet
> But there are your friends and your kind master.
> Bloom! Bloom, beautifully as in your native home
> Every spring.
> Your blossoms are a chain of friendship
> Between England and Japan.
> You are a speechless diplomat
> Oh *sakura*, *sakura*, my dear *sakura*.

24. Gambling with Success

One April day in the late 1920s, about the same time as Collingwood Ingram was trying to return his beloved *Taihaku* to Japan, another cherry tree in his garden at The Grange bloomed for the first time. Ingram was transfixed. The flowers were stunning. The single blossoms were smaller than *Taihaku*'s, but were perfectly proportioned and as white as pure snow. At first, Ingram didn't know the tree's lineage. But from the blossoms' peculiarities, he soon deduced that this unknown plant's parents were the large, white-flowered Ōshima cherry and the smaller Fuji (*Mame-zakura*) plant. As far as Ingram could tell, the tree was a natural hybrid that resulted from the seeds' parents being of different species.

Cherry trees hybridise easily when two cherries of different species or varieties come together. When they pollinate, a new type of cherry is produced. Whenever such a cherry appeared, Ingram put on his detective's hat, observed the colours and structure of the plant and attempted to work out its heritage.

The parents were often easy to pinpoint because, much like humans, most plants bore some resemblance to their mother and father. As this snow-white Ōshima–Fuji child was a new cherry variety to Ingram, he gave it a name that reflected his longtime interest in birds: *Umineko* or black-tailed gull. In full bloom, he said, the blossoms looked like an enormous flock of gulls nestling on top of a small tree. When he then exhibited the unusual blossoms at the 1928 Royal Horticultural Society's annual flower show in London, the judges awarded Ingram their Award of Garden Merit. Since then, *Umineko* cherries have often been planted as street trees in England, but have also become popular in parks around the world, from Kew Gardens in London to Queen Elizabeth Park in Vancouver.

Umineko's sudden appearance intrigued Ingram. If such a beautiful specimen could spontaneously appear through natural cross-pollination, perhaps he could create his own cherry hybrids? No one had ever deliberately hybridised cherries before, and the Japanese themselves did not do so until after the Second World War. During the Edo period, when more than 250 cherry varieties were created, trees were cultivated mainly by picking and planting the seeds of the most beautiful cherries. Only at the end of the nineteenth century was the artificial hybridisation technique successfully applied to other agricultural plants to improve farming methods.

Eager to become the first person in the world to hybridise cherries, Ingram sought advice from John Charles Williams, the owner of the Caerhays Castle Estate in Cornwall, who had become an expert hybridiser of rhododendrons. Failures, Williams told him, far exceeded successes, although the anticipation of success was a pleasure in itself. Ingram, never one to admit defeat, was intrigued by the possibilities. After all, a fabulous plant collection was one thing, but the creation of new varieties was on a higher plane – one on which only the fittest survived:

> Hybridising plants [is] not only the most inexpensive but
> also the most exciting form of gambling I know – a game
> of chance where the stakes concerned are no more than the
> common currency of every gardener – time and trouble.
>   Admittedly, the odds are very heavily against one. But what
> a thrill, what a joy it is when one does bring off a winner.

Cross-pollinating the cherries was a particularly refined and delicate process. First, Ingram grew in a pot the plant that he intended as the female parent. This he kept in a greenhouse away from other cherries, to prevent it being inadvertently pollinated by another tree. When the tree began to bud, but before the buds had fully bloomed, he would snip off the anthers – the tips of the stamen, or male organ, that hold the plant's pollen sacs – with a pair of small scissors or forceps. He waited for the plant's pistils, or female organs, to become ready for pollination, and then fertilised them with pollen taken from the tree that he wanted to be the male parent. This kind of pollination required meticulous planning and care. As different varieties of cherry bloomed at various times, the pollen from one parent had to be collected beforehand and carefully preserved until it was used.

'The captain would kneel in his greenhouse, scissors in hand, hybridising the cherry trees and looking really serious,' Ruth Tolhurst told me. We were sitting in the flower-filled cottage in the grounds of The Grange where she had been born eighty-nine years earlier. Tolhurst's father, Albert Stannard, had looked after Ingram's horses when the family lived at Westgate-on-Sea. After Ingram moved to The Grange, he asked Stannard to manage the dairy cattle on the farm next to the house in Benenden, which he had also bought. After marrying, Ruth left the cottage. She returned in 1972 after her mother died, to look after her father, and has remained ever since. 'I've been fond of gardening since I learned how to take care of flowers and trees from the captain when I was a child. He was always willing to pass on his knowledge. But I wasn't allowed to interrupt him when he was in the greenhouse. He concentrated so intensely on his work that you could feel the tension in the air.'

A large measure of patience, not to mention luck, was required to bring off a winner. After cross-pollinating two cherries, Ingram would sometimes have to wait for years after planting their seeds, before seeing a resulting tree. When the buds finally did open, the blossom was rarely the masterpiece he had imagined.

He had long raved about the tall, mountain-dwelling Sargent cherry, a wild species that he had seen near Mount Fuji. And he also loved the claret-tinted blossoms of the bell-flowered Taiwan cherry species, also known as *Kanhi-zakura*. Surely, he thought, a combination of the two would result in a hardy cherry tree with rich, ruby-pink blossoms?

For years, Ingram tried to mate these two favourites. Finally a single seed germinated. And flowered. And disappointed. His precious seedling produced a few miserably small flowers. Instead of being ruby-pink, they were a dirty shade of white. 'In short, what I had confidently hoped would be my greatest triumph turned out to be a tree fit only for the bonfire,' he wrote. The hybrids were a constant reminder 'of my all-too-often futile attempts to improve on Nature'.

In the early 1940s, forsaking Sargent cherry, Ingram decided to cross-pollinate two other wild species: Taiwan (*Kanhi-zakura*) and Fuji (*Mame-zakura*). Taiwan cherries thrived in the tropical climate of Japan's southern islands of Okinawa. Meanwhile, the hardy white-blossomed Fuji bloomed about 1,000 miles to the north, around the mountain for which it was named. To make the task more difficult, Taiwan cherries bloomed in February and Fuji in April. Ingram hoped to create a small but sturdy new flower, with deep-pink blossoms, out of the two distant and distinct species.

The only problem was that he didn't have any Taiwan cherries at the time. One place that did was the Temperate House at Kew Gardens. So there, late one February, Ingram shook the pollen from the Taiwan cherry's ripe anthers onto tissue paper, folded them carefully and placed them in a Thermos flask, with a pinch of calcium chloride at the bottom to absorb any humidity. By keeping the pollen dry and stored at an even temperature, he was able to preserve it for nine weeks until the late-flowering Fuji was ready to be fertilised. At last Taiwan and Fuji produced a beautiful offspring.

Ingram named his creation *Okame*, after a Japanese goddess of good fortune and mirth. Its flowers bloomed each March, at the midpoint between the blossoming of the Taiwan and Fuji cherries. Each tree was bedecked in countless little petals, like stars in the night sky. Each bloom was tiny and delicate, taking after the maternal Fuji cherry. But each was also tinted a light pink by the mix of the two parents' shades. Better still, the sepals that supported the petals were a deep, vibrant pink. Ingram said the flower would 'be appreciated by all who have an eye for elegance of form and unpretentious beauty'. He was ecstatic. 'The offspring of this union has more than fulfilled my expectations,' he wrote.

Heartened by this triumph, Ingram set about trying to create more hybrids. Another success was a hybrid of Taiwan cherry and Kurile cherry (a variant of the Japanese alpine cherry, *Takane-zakura* species), the latter being originally from the Kurile islands, north-east of Hokkaido, in the cold far north of the Japanese island chain. Natural cross-pollination of the two cherries would have been virtually impossible, because the trees grew thousands of miles apart. But with Ingram's midwifery, the result was a horticultural gem, which he named *Kursar.*

When the *Kursar* flowered in early March, it produced what Ingram called the richest-coloured cherry he had ever created. Both *Okame* and *Kursar* won Awards of Garden Merit at Royal Horticultural Society shows. These days the two varieties blossom each spring around the world.

Besides his treasured cherries, Ingram also hybridised other flowers. In all, he created more than thirty rhododendron hybrids, giving them names ranging from Thimble, Throstle and Timoshenko to Snow Leopard and Willy-Nilly. He also created the June-flowering Solstice azalea by crossing Japanese cultivars, and he produced a white saucer-shaped cistus hybrid with a yellow button-eye. This he named *Cistus* 'Pat', in honour of a young Benenden girl called Patricia Thoburn who had moved to the village at the age of nine and would soon become a close friend of Ingram.

25. A Fairy-Tale Garden

'When you entered the gates that led to The Grange, the sight of the blooming cherry blossoms would take your breath away. You felt like you were being lured in by fairies.'

Sitting in the spacious sitting room of Pympne Manor, a fifteenth-century house in Benenden, Patricia Thoburn closed her eyes and reminisced about the glories of Ingram's garden in the late 1930s. She had moved to the timber-framed house with her parents in 1936. Her father, Hugh, was a wealthy and successful land developer and banker, who had retired at the age of forty-five and had bought farms and cottages in and around the village. His passion was gardening, so it wasn't long before his path crossed with Ingram's. 'My father and the captain – that's what I always called him – only talked about one thing: plants,' Patricia recalled. Hugh was eager to transform part of his property into a spectacular garden, like Ingram's, and the men exchanged plants and horticultural knowledge.

On visits to The Grange, Ingram would show the Thoburns the flowers and trees that he had gathered on his regular plant-hunting expeditions during the 1920s and 1930s. There were hydrangeas from the slopes of the volcanic Mount Aso in Kyushu, collected in May 1926; more than 100 gladioli species that he had found on a 1927 trip to South Africa; a trove of pale- and dark-blue irises, Atlantic cedars and a semi-evergreen oak tree from a 1931 trip to Morocco. And a handsome gorse-like broom, subsequently named *Cytisus ingramii*, which was an unknown species until Ingram found it in north-west Spain in spring 1936, shortly before the outbreak of the Spanish Civil War.

Ingram brought back these plants' seeds, often in large, screw-topped sweet jars that he half-filled with damp, coarse sand atop a few lumps of charcoal. Others travelled home in small boxes or cigarette tins, occasionally as seedlings, which he would perpetually sprinkle with water. Some of the plants failed to survive at The Grange, but many lived to remind him of 'some dimly remembered scene or of a half-forgotten incident. Or, more poignantly, of some person or place I know I shall never see again.'

But to the Thoburns and most other visitors, all these plants were mere ancillaries to The Grange's main event: the cherries, blooming one after another, spring after spring. Walking the paths of The Grange at that time of year was like wandering through a magical symphony of colour and form, each variety being an instrument that brought its own unique timbre to the performance.

In mid-March, Ingram's *Okame* and *Kursar* varieties would herald the new season, prompting their creator to share his joy with his Benenden neighbours. Ruth Tolhurst recalled Ingram's passion whenever his 'children' blossomed: 'When a new tree flowered, he'd come to the cottage and bang on the door and shout: "Come on! Come on! The blossoms are out. Come and see them!" I would fly out to see them, abandoning the housework.'

Cherry trees in blossom along the drive at The Grange

Peter H. Kellett, a nursery owner who also lived near Benenden, would receive urgent phone calls from Ingram, insisting that he drive immediately to The Grange to witness the blooming of Ingram's hybrids. 'I had to climb in my van and rush over to the house, which took about fifteen minutes, to see the blooms,' he recalled. 'The captain wouldn't take no for an answer.'

Another early bloomer, usually in late March, was *Ito-zakura*, or weeping spring cherry, which bore single white-petalled flowers. About the same time the plentiful *Somei-yoshino* bloomed for about a week before the soft pinkish-white petals fluttered to the ground. Fuji cherries, for which Ingram had a particular weakness, followed. It's a 'sturdy, stocky little tree,' he wrote. 'Whether in sunshine or shade it is always happy and cheerful. Nothing seems to daunt its determination to flower, and each spring it completely clothes itself in snowy blossom.'

The peak blooming season came in April, when most varieties would bloom, including Ingram's all-time favourite, the large white-blossomed *Taihaku*, as well as *Yama-zakura* and Sargent cherry. These were followed by the fragrant semi-double pink *Sumizome*, which Ingram had imported from the Arnold Arboretum in Boston. *Kiku-zakura*, or chrysanthemum cherry, so named because the densely packed petals resembled their namesake, would reach full bloom in mid-May, accompanied by the shell-pink semi-double petals of *Hōrinji*, named after a Buddhist temple in Kyoto. Concluding the seasonal symphony near the end of May were several late bloomers such as *Shōgetsu* and *Fugenzō*. Then in the autumn, the cherry trees' leaves would ripen into golden reds, yellows and browns.

There were one or two exceptions to these seasonal rhythms, for as the trees' leaves started to fall with the approach of winter, the semi-double flowers of *Jūgatsu-zakura* (in Japanese, *jūgatsu* means the tenth month of the year) would bloom, weaving flashes of pinkish-white colour through the tree's slender branches. *Jūgatsu-zakura* cherries bloomed in April for a second time. Equally unusual was *Fudan-zakura*, also known as the Ever-flowering cherry. These would blossom as early as November and flower until April, and Ingram would take cuttings inside to display at Christmas. The seasonal cycle was complete.

In 1925 Ingram had forty classified types of cherry tree in his garden and twenty-nine more varieties that he could not identify. Four years later, when he published an RHS paper on cherries, he confidently classified fifty-nine different cherries at The Grange and said that many more unidentified trees were also present. And by the 1930s more than seventy different kinds of wild and cultivated cherries were growing at The Grange, and there were probably many more.

Ingram didn't like all these cherries, but he tolerated them at The Grange because they were part of his collection. He disdained upright trees with rigid branches; he despaired when unhealthy trees were stunted; and he had no time for *Chōji-zakura*, or Clove cherry, with its small white flowers and fast-falling petals, saying that he had the 'doubtful distinction' of introducing it to England in 1924. 'It cannot be a thing of beauty. There are many other species more worthy of garden room,' he wrote.

Another also-ran was *Miyama-zakura*, or Korean mountain cherry, which bore small white flowers and was, according to Ingram, 'rather a shy flower and therefore of little ornamental value'. His greatest wrath, however, he saved for a cherry that is now among the most popular in the West, *Kanzan*.

26. 'Obscene' *Kanzan*

R ead the brochure of almost any garden centre, nursery or horticultural business and you will probably find an avalanche of information about the large, double-petalled blooms of the *Kanzan* cherry. The Royal Horticultural Society praises its coppery-brown young foliage and purplish-pink flowers. The US Arbor Day Foundation calls it a 'splendid specimen' because of its stunning pink blossoms, upright vase-shaped branches and ease of cultivation on different kinds of soil. Collingwood Ingram had three words for it: Showy. Vulgar. Obscene.

Kanzan was among the first cherry varieties to be introduced to the British Isles in the late nineteenth century, and its vibrant colours attracted flower-lovers. Their admiration contrasted with the attitude of the Japanese, who, sniffed Ingram, 'are far too artistically minded to plant such a vulgar and garish variety for their own edification'. The Japanese, he said, echoing the words of a botanist called Mr Hamaguchi, whom he had met in Kyoto, 'often liken these gross forms to the painted women of the towns whereas they compare the quieter charm of *Yama-zakura* with the country girl'.

Kanzan's flower gave Ingram an impression of overt promiscuity. 'It flaunts its finery with nauseating frequency,' he wrote, his words dripping with disdain. 'The eye quickly becomes tired of the aggressive beauty of these cherries.' One spring, Patricia Thoburn recalled, she and her mother had been horrified to see Ingram throwing a healthy young

Kanzan sapling onto a bonfire at The Grange. 'Collin, you can't do that to a plant,' her mother had cried, causing Ingram to grab the tree from the flames and hand it to her without a word. More than seventy years later, Thoburn pointed out of the window of her home and laughed. In her garden the sapling was now a flourishing, thick-trunked, 40-foot-high tree.

As Ingram grew older, his distaste for this variety intensified, because tens of thousands of *Kanzan* trees had been planted alongside roads in southern England. He called this 'an inexcusable violation of the native scene – an act as unseemly and incongruous as would be the erection of a pagoda in the precincts of St Paul's Cathedral'.

In about 1970, recalled Peter Kellett, a *Kanzan* tree was planted in the grounds of Benenden School, a prestigious independent boarding school for girls. Founded in 1923, the school occupies part of a mansion whose former owners included the Earl of Cranbrook and Viscount Rothermere. Ingram's daughter, Certhia, and Certhia's own daughters, Veryan and Frances, attended the school, as indeed did Princess Anne, the Queen's daughter, in the 1960s.

When the *Kanzan* was planted, Ingram was about ninety years old and an influential figure within the village. Furious at the planting, he arranged a meeting with the headmistress and told her that such 'obscene cherry blossoms' had no place in a school for girls. 'They're prostitutes, blowsy and always showing off,' he told the headmistress. Shortly after that meeting, Ingram quietly asked Kellett, who was helping him propagate cherries in his old age, to 'transform' the *Kanzan* tree at the school by grafting it with Sargent cherry. Finally chivalry prevailed and the girls' honour was protected.

27. The Cherry Evangelist

Aside from the *Kanzan* cherry, Ingram was evangelical in his attempts to spread Japanese wild and cultivated cherries far and wide. He cared little about a person's status, gender or nationality, as long as they shared his enthusiasm for the collection and preservation of diverse plants. 'Every gardener is a plant hunter,' Ingram told members of the RHS. 'For the more sedentary this horticultural sport − and for most of us, a sport it certainly is − probably entails nothing more strenuous than the careful scrutiny of a number of trade catalogues. For the majority it usually takes the form of an interchange of cuttings, seeds or bulbs with some fellow gardening enthusiast. In short, it is a pastime that possesses all the glamour of a treasure hunt with a reasonably good prospect of a handsome reward at its end. And what a thrill of satisfaction that discovery will give one!'

Besides tapping experts in Japan and the United States, Ingram relied on an elite network of gardening enthusiasts in Britain to help build his collection and share information. They were all privileged landowners, mostly educated at Eton, Harrow or other public schools and then at either Oxford or Cambridge University.

The men belonged to an exclusive and largely unknown gentlemen's club for garden-owning 'amateurs', who actively cultivated and exchanged plants that had been recently introduced to Britain. The club, called the Garden Society, was formed in November 1920. All its members were fellows of the RHS and, at their twice-yearly dinner meetings, they would bring along plants to talk about and exchange.

Ingram wasn't admitted to the Garden Society until 1957, when he was seventy-six years old. Despite his wealth and social standing, he wasn't a major landowner, and it wasn't until the society invited a handful of 'professional' gardeners to join in the 1950s that Ingram got the nod. Yet he had long been friends with virtually all its members, corresponding with them, staying at their houses and occasionally travelling with them on plant-hunting expeditions.

In September 1927, for example, Ingram went plant hunting in South Africa for three months with Reginald Cory, the owner of Dyffryn Gardens in the Vale of Glamorgan, and Lawrence Johnston, creator of the gardens at Hidcote in Gloucestershire. The fourth member of the expedition was Sir George Taylor, a Scottish botanist who later became director of the Royal Botanic Gardens, Kew. Ingram's other Garden Society friends included Gerald Loder, the owner of Wakehurst Place garden in West Sussex; and Sir George Holford, who developed Westonbirt Arboretum near Tetbury in Gloucestershire. Another wealthy friend and Garden Society member was Major Lionel Nathan de Rothschild, a Conservative politician-banker, who sponsored plant-hunting expeditions to the Himalayas and memorably described himself as a 'banker by hobby, a gardener by profession'.

In sum, the Garden Society was an all-male *Who's Who* of gardening enthusiasts and experts, many of whose gardens remain among the best-known botanical institutions in the UK. Yet despite these gardens' popularity, the Garden Society itself is virtually unknown, even among horticulturalists. It has no apparent public presence, though its patron is Prince Charles. As patron, he succeeded his grandmother, Queen Elizabeth the Queen Mother, the only woman with entrée to this eminent club.

No amount of money, aristocratic connections or gardening expertise could open the Garden Society's doors to two of Ingram's closest fellow enthusiasts. Vita Sackville-West, the novelist and garden designer, moved in the 1930s to Sissinghurst Castle, a ten-minute drive from The Grange at Benenden. She and Ingram would occasionally visit one another to talk about and exchange plants. Meanwhile, Marion Cran, Britain's first woman gardening broadcaster, lived even closer to Ingram, having moved to Benenden at about the same time as he did, in 1919. Cran was a frequent visitor to The Grange, gathering 'scraps of knowledge' from Ingram for her broadcasts, books and magazines.

To further increase general interest in cherries, Ingram opened The Grange to the public one weekend each spring, as part of a charitable project called the National Garden Scheme. The money raised went to district nurses, in an initiative that traced its roots back to 1860. For almost forty years, from the late 1920s onwards and throughout the Second World War, the annual opening of The Grange was a major event in Benenden, attracting visitors from miles around to see what the local newspapers billed as Ingram's 'famous collection of 500 Japanese flowering cherries and rhododendrons'. At the time, private collections of ornamental cherries in Britain were rare, and Ingram's was the largest and most spectacular.

Ingram also encouraged commercial plant traders to visit The Grange as a way of distributing cherry varieties, offering scions of his trees to anyone who wanted them, free of charge. 'I get the impression he wasn't at all business-minded,' Nick Dunn, the third-generation owner of Frank P. Matthews Ltd, a plant wholesaler, told me. Dunn recalled that his father, Andrew, and his uncle visited The Grange with other nurserymen to discuss cherries with Ingram. It was a win–win situation. Ingram spread his trees, and the traders spread their businesses.

That company, Frank P. Matthews, now supplies ten thousand Japanese cherry trees a year to garden centres, having significantly increased their circulation throughout the UK. In the 1960s they started selling three popular varieties – *Ukon*, *Ichiyō* and *Amanogawa* – along with others that Ingram had either collected or named, such as *Umineko*, *Asano* and *Taihaku*. Other horticultural vendors, including Notcutts nursery of Woodbridge, Suffolk, one of the oldest plant companies in Britain, also started selling plants that originated at The Grange.

Ingram wrote prolifically about his cherry-blossom addiction. An historical overview of cherries appeared in 1934 in the *Illustrated London News*, at which Ingram's elder brother, Bruce, remained editor. Under the headline 'The Cult of the Cherry Blossom: The Japanese Emblem of Loyalty & Patriotism', Ingram explained how the beauty of the blossoms had permeated Japan's poetry, art and history for more than a thousand years. 'And there is absolutely no reason why we should not make our parks and cities as beautiful in spring as those of the Japanese Empire,' he concluded. He expanded on this theme in other magazines, noting the pros and cons of planting certain cherries in British soil.

Cherries thrived almost anywhere, as long as the soil wasn't too dry and sandy, or too wet and heavy, Ingram said. The perfect spot was a south-sloping bank overlooking a lake or pond, where the trees could obtain both moisture and sunshine. As for which specific type of cherry to buy, Ingram was circumspect: 'To ask a mother which of her children she prefers is a question she would be unable, or unwilling, to answer. So it is with Japanese flowering cherries. Like the brands of a well-known beverage, they are all good — but some are better than others.'

The best, he advised readers, were the wild cherries, such as *Yama-zakura*, Sargent cherry and *Edo-higan*. Among the cultivated varieties, he chose *Taihaku*, *Hokusai*, *Shirofugen* and *Shirotae* because of their vigour and large flowers. While he preferred the simple single-petal varieties, these were less popular in Britain than the more fancy double blossoms.

In all, Ingram introduced about fifty different kinds of cherry to Britain, including *Imose*, *Taoyame*, *Kokonoe-zakura*, *Asano* and *Chōji-zakura*, or Clove cherry. He exhibited these and other plants at the prestigious flower shows organised by the Royal Horticultural Society. Between 1928 and 1966, Ingram's cherries won fifteen Awards of Garden Merit and numerous other prizes.

By the end of the 1930s Ingram had become a familiar name in horticultural circles, acknowledged as an expert in the taxonomy, collection, propagation and hybridisation of cherries. Now, to most friends and contacts, he was no longer 'Captain' Collingwood Ingram, the compass-adjusting cognoscente of the First World War. He was Collingwood 'Cherry' Ingram, or more simply 'Cherry', and the most prominent and celebrated resident in the tiny Kent village of Benenden.

28. Darwin Versus the Church

S trolling through the village of Benenden on a summery Sunday evening during the 1930s, a casual visitor would have detected little change in atmosphere from a century earlier. Cricket was being played on the triangular village green, as it had been since at least 1798. On one side of The Green, thirsty cricketers and locals sipped pints of bitter under the carved beams of the Bull Inn, the local pub. And at the top of the neatly mown triangle, churchgoers made their way to St George's parish church, close to which stood The Grange, home of the Ingram family.

Yet Benenden by no means reflected the rest of Great Britain, which was a land divided – by class, income and geography. While the British economy had struggled in the 1920s, conditions worsened in the 1930s, after the Wall Street Crash of 1929 caused exports to plummet and unemployment to soar. Tens of thousands of workers in the shipbuilding, mining and steel industries lost their jobs, mostly in northern England. Amid all the poverty and distress, some parts of Britain remained largely unaffected, including Benenden.

To be sure, the village had changed dramatically since the First World War, when the owner of the original estate, Baron Rothermere, had sold off his vast landholdings. Many Benenden farmers and agricultural workers lost their livelihoods. But wealthy Londoners, attracted by the ease with which they could travel to the capital by train or car, flooded into the village, turning yeoman houses into single-family dwellings. Others bought holiday or summer homes as a weekend retreat. Both groups included large numbers of former military officers, among them Ingram.

The presence of so many newcomers, as well as changes in British economic conditions, altered Benenden's character, although not its overall population. Before the war a majority of the village's 1,400 residents had worked on the land. Afterwards, when agricultural jobs declined because of the Depression, many farmhands and their families were hired as gardeners, drivers and housemaids by the wealthy arrivals. Others worked further afield or at the newly opened private Benenden girls' school.

At the core of the Benenden community in the 1920s and 1930s was St George's, a bastion of Church of England religiosity named after England's patron saint. There the residents interacted at Sunday worship, and at charity events and activities organised by the church's Ladies' Society and Young Men's Association. Attendance at church was a given for a large number of villagers, although Benenden's most renowned inhabitant was always absent. While Florence

Ingram and her children always went to church, slipping through The Grange's rear gate near the churchyard, Collingwood enjoyed his own spiritual compact with nature in the privacy of his flower-filled garden.

Ingram lacked religious belief and rarely attended church, no matter how much the rectors of St George's tried to persuade him. 'My father and Ingram were very close, as were both our families,' Anthony Price, the son of St Georges' vicar, Jessop Price, told me. 'They would sometimes debate Christianity and the theory of evolution. But when my father asked Ingram to attend Sunday services, the captain would always insist upon the primacy of evolutionary descent. I think he was trying to convert my father.'

A naturalist since his earliest days, Ingram subscribed to the mid-nineteenth century theories of Charles Darwin, Alfred Russel Wallace, Sir Charles Lyell and others about natural selection. He believed that plants had enough intelligence to aid their survival, and a sensitivity that was 'not entirely unlike the consciousness in animals'. Ingram realised that this sounded impossible to most people, but he was convinced that a plant's root system had an innate and uncanny power to seek the underground moisture it needed to survive.

For Ingram, the 1930s – the sixth decade of his life – were a time of contentment and contemplation in Benenden. He was active in village life: he was president of the Benenden Football Club and he judged gymkhana competitions each year for charity. Florence, meanwhile, was a soprano stalwart of the Benenden Choral Society, specialising in bird-related songs such as 'The Cuckoo', 'Sing, Joyous Bird' and 'I Hear a Thrush at Eve'.

A self-portrait by Ingram, 1930s

Ingram punctuated his activities with months of well-planned travel to collect plants, watch birds and see the world. In the 1930s he visited India and the West Indies, the Azores and the Falklands, New York and New Zealand. But as the 1930s drew to a close, the political mood around the world became increasingly sombre. After twenty years of uneasy peace in Europe, Adolf Hitler had grabbed power in Germany and crushed any opposition. In 1938 Germany annexed Austria and occupied Czechoslovakia.

In early July 1939, even when Ingram believed that 'war with Germany seemed well-nigh inevitable', he travelled to Iceland for two weeks and drove to the island's north-western corner to go salmon-fishing and to visit the breeding grounds of dunlin, golden plover and whimbrel.

Six weeks after Ingram returned to England, Germany invaded Poland; and two days later, on 3 September 1939, Britain declared war on Germany. The first seven months were a so-called 'phoney war', during which the British were not involved in the fighting. But by the spring of 1940 the lives of most Benenden residents, including Collingwood and Florence Ingram, changed dramatically, as the reality of a possible German invasion started to hit home.

29. The Sounds of War

Saturday, 27 April 1940 was a magical day for the Ingram family. Collingwood and Florence's daughter, Certhia, aged twenty-three, married Gerald Harden, a Benenden farmer. After the ceremony in St George's Church the newly-weds and their guests gathered for a reception in the sunlit gardens of The Grange.

It was the height of the cherry-blossom season, and an abundance of pink and white blooms greeted the wedding party. 'The cherry blossoms danced around the pure white of Certhia's wedding dress. It was a magical sight, just like you'd see in a film,' Patricia Thoburn recalled. After the wedding, Certhia and Gerald set up a new home in Benenden.

It would be more than five years before such tranquillity would return to Benenden and the rest of the British Isles. Two weeks before the wedding, Germany had invaded Norway and Denmark. Seventeen days after the nuptials, on 13 May, the new prime minister of an all-party coalition government, Winston Churchill, outlined to the House of Commons the difficulties that lay ahead. 'We have before us an ordeal of the most grievous kind,' he told Members of Parliament. 'We have before us many, many long months of struggle and of suffering.' That same day the Nazis invaded France, Belgium, Luxembourg and the Netherlands.

By then, petrol and popular food items, such as bacon, butter and sugar, were rationed and the British armed forces had conscripted more than one million young men. Ingram's three sons — Ivor, Mervyn and Alastair — all joined up. The eldest, Ivor, became a flight lieutenant in the RAF Volunteer Reserve. Mervyn was a doctor in the Royal Army Medical Corps, attached to the 1st Field Regiment (Royal Artillery) and later the airborne division. Alastair was a regular soldier in the Royal Artillery, posted to Hong Kong, where he met his future wife, Daphne.

Residents of Benenden felt particularly vulnerable, since the village was less than 15 miles from the English Channel, which divided Britain from the European continent. There, thousands of German troops were massing in preparation for a possible invasion.

It was time for action. In a radio broadcast on 14 May 1940, Anthony Eden, the Secretary of State for War, asked for men aged seventeen to sixty-five to form Local Defence Volunteer forces throughout the country. Ingram was among the first of about thirty men in Benenden to sign up. As one of the few veterans of the First World War, the 59-year-old captain was appointed commanding officer of the unit. The military provided the unit with bare necessities: twenty rifles, 200 rounds of ammunition, a few torches, 'LDV' armbands and an emergency telephone authorisation card for calling Ingram at 'Benenden 2115'.

It was in this febrile atmosphere, on the evening of Sunday, 26 May 1940, as the deadly sounds of war echoed across the English Channel, that Ingram became transfixed by the liquid trills of one of his favourite songbirds – the nightingale. Its song cut through the gloom and pierced his soul.

Ingram had loved nightingales since childhood. As a small boy in the 1880s he had visited the village of Bellagio, on the shores of Lake Como in Italy, with his parents. His most vivid recollection of that trip was the extraordinary number of nightingales that could be heard singing around the resort. 'From every side and all night long, the air reverberated with their song,' he wrote later.

Three decades later, on the evening of 6 July 1917, in radically different circumstances, Ingram had shut out the nearby war by listening to a nightingale in a forest near Saint-Omer in northern France. 'Last night was one in a thousand – a full moon riding in a mellow summer sky and not a breath of wind to stir the soft scented air,' he wrote in his diary the following day.

In particular, Ingram loved the romantic idea that the nightingales he listened to in Benenden had travelled all the way from tropical Africa. On summer evenings, he said, he could sometimes not only hear, but actually feel their song, because 'the sound waves emanating from the bird's loud throbbing notes would beat a soft, faintly perceptible tattoo upon one's eardrums'.

Now, in May 1940, as Ingram patrolled near Benenden, a nightingale's song once again prompted him to reflect on the munificence of nature and the evanescence of life. His daughter Certhia's wedding reception just one month earlier, beneath the cherry blossoms at The Grange, seemed like a lifetime ago.

Britain's future appeared bleak. As head of the Home Guard unit, Ingram divided his men into north and south Benenden patrols and set up a nightly rota. Confront all persons who approach your posts, and any other person whose genuineness you may suspect, Ingram instructed his volunteers. Satisfy yourselves that they are harmless citizens, before allowing strangers to pass. 'The challenge will be "Halt! Who Goes There!"' he shouted.

Ingram's diary entry on 26 May summed up the villagers' gloom:

> The end of all happiness seems to be approaching. Dark destiny thunders loudly and ominously in the air all day and all night. The dull crumpling roar of distant artillery throbs increasingly in our ears like the angry, muffled roar of some all-destroying monster.
>
> There was no joy to be found here. Only the wail of a little owl interrupted the thunder of the guns. All along the southern skyline, their flashes flickered like summer lightning, a lightning that spelt death and destruction across that narrow strip of water.

That day had seen the start of Operation Dynamo: the evacuation of British Expeditionary Forces and other Allied troops from the beaches at Dunkirk. Over the following ten days more than 335,000 soldiers were ferried to Britain in a makeshift flotilla of boats that became known as the 'miracle of Dunkirk'.

A special prayer service was held at St George's Church in Benenden that evening, as it was in hundreds of religious institutions throughout Britain. In an overcrowded Westminster Abbey, King George VI, Prime Minister Winston Churchill and the entire Cabinet prayed with the Archbishop of Canterbury for soldiers 'in dire peril in France'. In Washington D.C., President Franklin D. Roosevelt broadcast a Fireside Chat to warn Americans of the 'approaching storm'. Many believed that the situation was the direst England had faced since Napoleon Bonaparte had sought to invade the country in 1805.

Yet as Ingram drove through the Kent countryside that night, the piercing song of a nightingale cut through the cool spring air.

> For a brief space of time, this awesome, nerve-wracking sound [of war] was completely drowned by the ecstatic outpouring of a nightingale. Never have I heard this bird sing more loudly or more exquisitely. The powerful musical notes filled the night with a resonant melody that was like an epitome of passionate joy.
>
> I stopped my car and lingered awhile, listening to this great, heartening voice, and for a brief, for a very brief, space forgot all else.

It was a moment of transcendental tranquillity. Nature humbled Ingram; it was his faith, and the nightingale's cry that evening reinforced his belief that nature's voice would always make itself heard, even at the darkest of times. The sound would sustain and invigorate him that summer, as the Battle of Britain raged over the skies of Kent and over his beloved cherry trees.

PART FIVE

FALLING BLOSSOMS

30. Cherry Blossom Brothers

Six-year-old Akiko Mitani stood to attention, hands rigid by her side, head held high, eyes straight ahead, and chanted the sentences that were among the first she had ever learned by heart. Her forty or so classmates at a primary school in Kameyama City, 50 miles east of Kyoto, joined in the complicated recitation of the imperial order. It was April 1937, the beginning of the school year. Outside, in the schoolyard, the cherry trees were in full blossom.

'Our Imperial Ancestors have founded our Empire on a basis broad and everlasting and have deeply and firmly implanted virtue.

'Ye, our subjects, be filial to your parents, affectionate to your brothers and sisters; as husbands and wives be harmonious, as friends true; bear yourself in modesty and moderation; extend your benevolence to all; pursue learning and cultivate arts, and thereby develop intellectual faculties and perfect moral powers.'

Akiko took a deep breath and continued. She didn't understand most of the words, which were written in an archaic form of Japanese that no one spoke any longer.

In a similar classroom in Okayama, 250 miles west of Kameyama, a six-year-old boy, Hiroyoshi Abe, was intoning the same words at almost the same time of day. Like Akiko's school teachers, Hiroyoshi's instructors continually reminded him that it was the duty of every proud Japanese to be modest, helpful, affectionate and loyal — to one's friends, family and, above all, the Emperor.

'Advance the public good and promote common interests, always respect the Constitution and observe the laws; should emergency arise, offer yourself courageously to the State; and thus guard and maintain the prosperity of Our Imperial Throne coeval with heaven and earth.

'So shall ye not only be our good and faithful subjects, but render illustrious the best traditions of your forefathers. The Way here set forth is indeed the teaching bequeathed by our Imperial Ancestors, to be observed alike by their descendants and the subjects, infallible for all ages and true in all places.'

In their respective classrooms, Akiko and Hiroyoshi — my parents — breathed out slowly at the end of the reading. Each morning before classes right across the country, every pupil recited by heart the *Kyōiku Chokugo*, the Imperial Rescript on Education that had been issued by the Meiji emperor in October 1890. Even today, my mother and father can recite the entire decree without error or prompting.

As well as swearing allegiance to the emperor day after day, Mama and Papa also learned about their imperial heritage as his 'children'. Even on *Kigensetsu* (Empire Day) and other national holidays, they had to attend school and participate in imperial ceremonies. History lessons began with creation myths that stretched back to the sun goddess Amaterasu and the first emperor, Jinmu.

My parents were both born in October 1931 at a time when, politically, Japan was drifting dangerously to the right and the military was tightening its grip over the civilian government. They came into the world just days after Japanese soldiers in north-east China staged an event known as the Manchurian Incident, which led to Japan's invasion of the resource-rich Chinese province of Manchuria and the establishment in 1932 of the puppet state of Manchukuo. Meanwhile, on 15 May 1932, nationalists attempted a *coup d'état* against the government, which resulted in the assassination in Tokyo of Prime Minister Tsuyoshi Inukai. The nationalists also targeted for assassination the Hollywood film star Charlie Chaplin, as a way to provoke the United States. Chaplin was in Japan to promote his movie, *City Lights*. But when the assassins stormed a VIP reception at the prime minister's residence, Chaplin wasn't there. He had gone with Inukai's son to a sumo wrestling match.

Japan's land grab in Manchuria so incensed the West that in 1933 the League of Nations recommended that Japan restore Chinese sovereignty in Manchuria. Japan refused and withdrew from the organisation. In 1936 Germany and Japan signed an anti-communist agreement called the Anti-Comintern Pact. The following year the Second Sino-Japanese War began, with Beijing, Shanghai and Nanjing falling to the Japanese in quick succession.

Japan's nationalists were jubilant at their nation's successes in China. Throughout the 1930s they had been advocating an 'Asia for Asiatics' — a region free and independent from the West. At the time the UK ruled India, Burma, Malaya and Hong Kong; France occupied the east of the Indochina peninsula; the US controlled the Philippines; the Netherlands ruled Indonesia; and Macau was a Portuguese colony. Much of China was nominally ruled by the Kuomintang, although the Western powers maintained extraterritorial status in many cities, including Beijing and Shanghai. That left only Thailand

and Japan, which had ruled Korea since 1910, as Asian nations free of a Western presence. In 1940, on the purported 2,600-year anniversary of Japan's founding, the nation's foreign minister called for the establishment of a Greater East Asia Co-Prosperity Sphere to unite the region politically and economically.

As young children, of course, my parents were unaware of the military's growing influence and control over Japan. But they both remembered the excitement in December 1941, when they were ten years old, after Japan attacked Pearl Harbor in Hawaii and declared war on the almighty United States of America and its allies.

One chilly evening in November 2016, Mama, Papa and I sat in my father's cluttered room in a care home in Ibaraki Prefecture, an hour north of Tokyo, and my parents continued their stories. Papa, now eighty-five, had just returned from four weeks in hospital after a fall. His memory was a little blurry, but nothing could erase the lessons he had rote-learned or the songs he had sung in the late 1930s and early 1940s.

In those early years of his life, everyone in Japan worshipped the emperor as a living god, and the cherry blossom was among the military's most powerful symbols. Of all the words in the Imperial Rescript on Education that Japanese children recited, the most important phrase was 'Offer yourself courageously to the State'. In other words, be ready to die for the emperor. Indeed, all soldiers were told that they must be prepared to fall and die in the same way that the cherry petals fell and died, after a short but lustrous life.

'Do you remember any of the military songs you used to sing during the war?' I asked my father.

A temple courtyard in Kyoto, tree described as *Prunus subhirtella* 'Autumnalis'. Photographs from Ingram's 1926 trip to Japan

Described by Ingram as '*Yoshino* cherry in Uji',
the tree is thought to be *Somei-yoshino*

At a temple gate in Ishiyama, Shiga Prefecture.
The tree is described as *Prunus mumé* (plum tree)

Kiyomizu temple in Kyoto

Fugenzō at Nikko

The weeping cherry tree at
Daigoji temple in Kyoto

An encounter in Ishiyama,
Shiga Prefecture

In the Yoshino mountains

He nodded and softly began singing a song called *Hohei no Uta* ('Song of the Infantry'):

The colour of a soldier's collar is
Just like many sprigs of cherry blossom
Blown in a Yoshino storm
Born as a son of the Yamato race
Fall as flowers on the front line.

The infantry soldiers' collar was scarlet, the same colour as some cherry petals. And Yoshino was a popular cherry-viewing destination, which Collingwood Ingram had visited in 1926. The 'Yamato race' was a nationalistic term, used mostly to distinguish settlers of mainland Japan from other races.

Papa recalled the cherry trees at his secondary school, especially one that seemed to stand guard over the pupils at the entrance. 'We marched around the school playgrounds for hours, sang military songs and did some military training,' he told me. 'The teachers put objects on the ground and we had to crawl through them.'

Papa's secondary school, which he entered at the age of twelve, was the best in the region, from where the brightest boys were expected to move on to the elite Imperial Japanese Naval Academy in Hiroshima Prefecture, or to related academies in Tokyo or Kyoto. Papa was top of the class academically.

'The problem was that I wasn't very sporty and I was slow,' he continued, 'so the teachers often hit me in the face. We all knew that we'd eventually have to join the military and fight.'

Papa wanted to join a naval institute. It was more appealing than the army. He and his friends all devoured a best-selling book, *Kaigun*, about one of the nine young naval soldiers who had died in the December 1941 attack on Pearl Harbor and had become national heroes.

'The way that navy officers saluted was much smarter than the army officers,' Papa laughed, raising his right hand to his cheek, rather than to the top of his head, like the army salute. Closing his eyes, he started singing again, this time a verse from a popular wartime ditty, 'Cherry Blossom Brothers' (*Dōki no Sakura*). In later life, Papa, a liberal journalist and an academic, had been fiercely against the war and had reviled the emperor-oriented ideology that had caused millions of deaths. Yet the words had lingered and another verse came tumbling out, as though it was 1942:

> You and I are cherry blossom brothers
> Even if we fall separately on different battlefields
> We'll meet again at Yasukuni Shrine
> As cherry blossoms, in the treetops in spring.

The Imperial Shrine of Yasukuni, to which the song referred, was a state-run Shinto shrine in central Tokyo that played a key role during the Pacific War – as the Second World War is known in Japan – as a symbol of dedication to the emperor. It had been built by the Meiji government in 1869, initially to commemorate those who died in the Boshin War of 1868–9, fighting for the emperor to overthrow the shogunate. After the military took control of the shrine in 1872, the war dead from the Sino-Japanese War, the Russo-Japanese War and the First World War had all been enshrined there, with their names and other personal details recorded. Over the years the shrine increasingly became an ideological institution to glorify the emperor and the

Empire of Japan. During the Pacific War more than two million dead Japanese soldiers were honoured at the shrine as 'gods of war', and the Japanese people worshipped them there. The soldiers were told that if they died for the emperor, they would live on as cherry blossoms within the shrine's grounds, where hundreds of trees had been planted. The majority of trees were of the *Somei-yoshino* variety.

'Falling was everything,' Papa continued. 'Falling for the emperor, graciously and bravely. We were told this, day after day after day. We needed to uphold the *Yamato damashii* [Japanese spirit] like samurai and then die like cherries. To talk about living was taboo. We were living to fall. It wasn't logical, but we were educated to think like that. No one questioned it.'

The symbolism of a beautiful flower that opened and then quickly shed its petals, reflecting the fleeting nature of life, became a key aspect of the narrative that Japanese children were taught from the day they started elementary school at the age of six. While the imperial family's official emblem was a sixteen-petalled chrysanthemum, the cherry was the people's flower. All schools in Japan followed the same curriculum, which was set by the Department of Education. All nursery school children sang:

Yama-zakura, Yama-zakura,
Release your fragrance
Even after falling, for the Emperor.

Meanwhile, older pupils read a textbook, first published in 1900, that extolled the cherry blossom's virtues and the tree's links to the emperor:

Japan is a small country on the map, but it is a great country. One needs to be proud to be born Japanese because our country is known for its deep affection for the Emperor and one's parents. The *botan* [peony] is the national flower of China and Westerners adore roses. But the Japanese love cherry blossoms' transparency and purity. Our hearts should be as transparent as the petals of a cherry blossom. Without this, one is not truly Japanese.

To reinforce their allegiance to the emperor, all Japanese school pupils were required to gather about once a week in front of a shrine. At my father's school, which was situated on a hill, the shrine was built outside near the top of the slope, in a commanding position. No one made a sound when the school principal opened the shrine's wooden doors, my father remembered. Behind the doors was the *Goshin'ei*, a sacred portrait of the Emperor and Empress of Japan. Sliding their hands down their thin legs to steady themselves, the small boys would then bow deeply in prayer and reverence to the living god.

Because these portraits were considered sacred, many schools posted 24-hour guards to protect them. Some school principals even died during the war trying to save the *Goshin'ei*, when their schools burned down after US bombing raids. My father's school was completely destroyed on 29 June 1945, after 140 US B-29 bombers dropped their munitions on the city of Okayama — the city where my father grew up — killing 1,737 citizens. But the shrine at the top of the hill escaped the bombings.

'I walked to school the morning after the bombings,' Papa said quietly. 'There were many dead bodies lying on the roads.' He paused and narrowed his eyes, reliving the images and shaking his head. 'When I reached the school, I saw the deputy head standing in front of the shrine, amidst the rubble, praying. Then several of my schoolmates showed up and we all helped the deputy head clear the area around the shrine.'

Mama joined in the conversation, as Papa flagged. 'Girls were also taught to be brave, Nao-chan,' she said, using a familiar term of endearment. 'Towards the end of the war, when everyone said that the Americans would soon be landing, we were given bamboo sticks at school and practised stabbing straw "American" dolls. We all believed, until the very end, that the emperor was a god. And then, as soon as the war was over, we were told that he wasn't. I was just thirteen, so it was very confusing.'

31. Flowers of Mass Destruction

Collingwood Ingram would have been appalled at my parents' recollections, particularly had he known the extent to which the government had subverted the symbolism of his beloved cherry blossoms. Ingram rarely mentioned or alluded to politics in his writing, although he was clearly disturbed during his 1926 visit to Japan by the direction in which the country was heading. He had always romanticised the 214-year *Sakoku* era before 1853, during which the feudal *daimyō* lords had cultivated hundreds of

cherry varieties, and he detested Japan's rampant industrialisation in the late nineteenth and twentieth centuries.
'Commerce and militarism are today paramount in the
people's thoughts, and the delightful old-world feudalism of
the previous generation is but a faded memory,' he wrote
in 1929, in a rare reference to the political winds blowing
through Japanese society.

It was ironic, an unnamed Japanese diplomat told Ingram in
the late 1920s, that during the *Sakoku* era, when the Japanese
people were 'the most artistic and aesthetic race in the world, we
were called barbarians. [But] now that we have become a power
and have acquired the means of killing our fellow men, we are
called a civilised people.'

Writing in 1936 about the 'sooty ugliness' of the port city
of Kobe, for instance, Ingram had been scathing about the
'unequivocal sacrifice of beauty to the god of Mammon ...
And for what purpose? To fill the government treasury that
the nation might become a world power? Or that it might be
in a position to slaughter, or to dictate terms to, its weaker
neighbours?'

By then, Ingram was clearly disillusioned with Japan.
Yet throughout the 1930s and 1940s — years in which Japanese flowering cherries dominated his life — he never again
mentioned Japan's march towards, and descent into, war. He
received occasional letters from Japanese cherry experts in the
1930s about certain varieties. But his own correspondence to
those men is missing. In any case it is unlikely that he would
have mentioned, let alone criticised, Japanese politics in his
letters to Japanese friends. It would have been too dangerous
for them.

Ingram would scarcely have imagined that the cherry blossom would become so closely linked to the Japanese military. Nor could he have dreamed that these ties would eventually affect his immediate family. The fact was that, in little more than a generation, Japan's leaders had quietly and imperceptibly transformed the symbolic meaning of the cherry blossom – flowers of peace for more than 2,000 years – into flowers of mass destruction. Never before had one flower's symbolism changed so rapidly. As the emphasis on what the cherries stood for differed over the years, the shift from 'living' to 'falling' went largely unnoticed, according to Emiko Ohnuki-Tierney, an anthropologist at the University of Wisconsin.

That was because cherry blossoms represented conflicting images of life and death to the Japanese throughout their history. The blossoms were primarily a symbol of life, cheerfulness, vigour and peace, but in Japanese literature they were also associated with fleeting beauty and the transience of existence, because of their short life. This perception of impermanence evoked images of mortality. At the same time, however, cherries symbolised rebirth, because they blossomed year after year. Since all these images coexisted in people's minds, the Japanese population did not notice when the flower's emphasis shifted from life to death.

The metamorphosis of the cherry's image began quietly soon after 1868, when the reforming Meiji government replaced the anachronistic Tokugawa shogunate. At the time the image-change was an almost insignificant part of the new government's ploy to thrust Japan into a modernised future by honouring one man, who had been almost forgotten by the general public – the emperor.

32. Emperor Worship

The term 'Meiji Restoration' does not do justice to the dramatic events in the late 1860s and 1870s that signalled the end of Japan's feudal age. The restoration was nothing less than a revolution, a historic change that totally upended the status quo. Modern Western legislative, administrative and judicial systems had evolved over several hundred years. The Meiji government wanted to replicate these in several hundred days. By and large they succeeded, and four political slogans of the age left no doubt about the government's goals: 'Catch Up with the West'; 'Wealth, Strength, Military Power'; 'Get Out of Asia, Enter the West'; and 'Civilisation and Enlightenment'.

In quick succession the Meiji government abolished the class system, made education compulsory for all children from the age of six, ordered military conscription for the first time and instituted a national tax system. Meanwhile, the feudal domains were re-designated as prefectures, which were similar to English counties and US states, and the *daimyō* lords who ruled these fiefdoms were replaced by governors sent from Tokyo.

But the Meiji leaders faced a dilemma. How could they unite, emotionally and spiritually, thirty-four million people who had no sense of belonging to a 'nation'? During the Edo period, from 1603 to 1868, everyone belonged to his own domain and was beholden to one of the 270 or so *daimyō*. No one called himself or herself 'Japanese'. But now, in case of an emergency, the government would need to convince millions of ordinary people to take up arms. Without the sword-wielding samurai to protect them, gun-toting farmers and their urban brothers would need to defend the country.

The Meiji leaders, mostly men from western and southern Japan, had other priorities. Under pressure to improve Japan's global status, they started negotiating with the West to rectify the unequal treaties that the shogunate had been forced to sign after the Black Ships arrived from the United States in 1853. One cause of indignation was that Westerners who lived in Japan did not have to abide by Japanese laws and could not be tried in Japanese courts if they committed crimes. Another was that Japan was not allowed to set tariffs on Western imports, which made it difficult for some industries to modernise and compete. Not surprisingly, the West pushed back against Japan's requests, saying the country still lacked a constitutional political and legal system. That rebuff sparked the Meiji leaders into action.

In 1882 a Japanese delegation led by Hirobumi Itō, who later became Japan's first prime minister, travelled to Europe on an eighteen-month trip to the newly unified Germany, Austria–Hungary, England and Belgium to learn about the region's constitutional systems. What was the monarch's role? How did parliament work? Who should be given the right to vote? On their return to Japan, Itō eventually decided to model the nation's constitution on that of Germany.

Japan needed a clear 'axis' or foundation that would function in the same way as the Christianity that underpinned Western culture, Itō concluded. While he was in Berlin, a German politician, Rudolf von Gneist, had suggested that Japan should make Buddhism the guiding force in its constitution. Itō disagreed, since Buddhism had come to Japan via China in the sixth century. Other religions — Confucianism, Taoism and

Christianity — also came from abroad. The only domestic faith was Shintoism, an animistic ethnic religion in which emperors had long played a role, conducting rituals around the time of the rice harvest.

Indeed, Itō realised that the sole element in Japanese society that had remained unchanged for more than 1,000 years was the imperial system. The Japanese monarchy was different from its Western counterparts, in that the throne had passed from one emperor to another without a single break in the bloodline. Itō concluded that a combination of the imperial system with the indigenous Shinto religion would give Japan a powerful spiritual backbone.

To support this idea, and to implant loyalty and devotion to the emperor among the general public, the Meiji leaders promoted three powerful concepts. The first was the *bushi*, or samurai, philosophy of moral ethics. The second was the nationalistic concept of *Yamato damashii*, often translated as the 'Japanese spirit'. And the third was the symbolism of the cherry blossom. Together, these eventually formed what I call the 'cherry ideology' — a tool with which the government effectively indoctrinated and controlled its people in the early twentieth century.

At the heart of this ideology lay the idea that the emperor was divine. This concept was originally created by a seventh-century emperor who was intent on burnishing his legitimacy. Emperor Tenmu, who reigned from AD 673 to 686, commissioned a written compilation of Japan's history, called the *Kojiki*. According to this document, the first Emperor of Japan in 660 BC, Jinmu, was a direct descendant of the sun goddess Amaterasu. This story was a contrived myth and a way for Emperor Tenmu and the imperial family to forge a distinctive dynastic identity for the Yamato people, who made up most of the population from the fourth century onwards.

In the eighth century, when Japan established its first centralised governance system in the city of Nara, the emperor had initially ruled as both a religious and a political figure. But as political power shifted gradually to the nobles who worked for him, he increasingly became a ceremonial figurehead. After 1192 samurai took over the role of the nobles, and political power shifted completely to the most powerful military commander, the shōgun. While the emperor's court remained in Kyoto, his role was confined to nominally appointing successive shōguns and performing rice-harvest rituals within the court. After Ieyasu Tokugawa united the country and set up his shogunate in Edo in 1603, he stripped even more power from the emperor by decreeing that his primary duty should be 'academic pursuits'. In essence, the emperor became almost invisible and inconsequential, although his ceremonial role never ceased to exist.

Almost three centuries later, when the Meiji government deliberately reinvented the imperial story, the emperor was no longer an irrelevance. Now he was cast as a 'divine father figure', with a role newly enshrined in a constitution that became the foundation of Japan as a modern nation. Issued in February 1889, the Meiji Constitution was well received in the West as it enabled the government to set up a parliament and electoral system. The new structure was designed to give the Japanese a sense of purpose, obligation and patriotism by putting the emperor at its apex while the people served as his *shinmin*, or 'subjects'. Alongside the constitution, the Shinto religion was recast as a state ideology.

'The Constitution had two conflicting principles,' the 100-year-old Kiyoko Takeda, Emeritus Professor of Modern Japanese Intellectual History at the International Christian University in Tokyo, told me, when I spoke to her in June 2017. She called this the 'two-horse carriage' system. The constitution was deliberately written to please two groups that had completely opposite views. One group consisted of Japanese liberal intellectuals and Western leaders who wanted to see Japan as a constitutional monarchy, she said. That horse veered to the left, carrying the message that the emperor was answerable to the constitution. The other group consisted of Japanese nationalists who believed that the emperor transcended the constitution. That horse veered to the right, with the message that the emperor was Almighty, Professor Takeda said. The government believed it could keep these two horses in balance, and the constitutional carriage on track. While the vast majority of the population did not have any opinion about the emperor, the Meiji leaders sought to convince them to revere him as a divine figure.

In the carriage behind the horses sat the prime minister, who was holding the reins and trying to maintain the carriage's balance. At first the horses were equally strong and the carriage ran smoothly. The left-leaning 'horse' started to prevail during the reign of Emperor Taishō from 1912 to 1926, a period known as the 'Taishō democracy', when ideas of human equality and freedom spread. Later the carriage gradually tipped in the opposite direction. By the 1930s it was speeding along the path towards totalitarianism and fascism, in a way the Meiji leaders could never have foreseen half a century earlier.

Alongside the new constitution, the Meiji leaders promoted *bushidō* – the 'way of the samurai' – as a core ethical standard for the nation (*Bushi* means 'samurai', while *Dō* means 'the way'). *Bushidō*'s values included loyalty, honour, courtesy, courage and chivalry. These standards, all designed to build a samurai's character, were repurposed for the new age and became cultural principles that embodied the *Yamato damashii*.

Inazō Nitobe, the author of *Bushidō: The Soul of Japan* in 1900, compared *bushidō* to a bright star, offering a moral path for all Japanese. No longer did these morals apply only to the samurai. They applied to everyone, as the supreme ethical cord linking Japanese nationals. And no longer did the *bushidō* philosophy signify loyalty only to one's *daimyō*. By the late nineteenth century it meant loyalty to the emperor.

In 1882 the *bushi* concept became part of the official code of military ethics, which all soldiers and sailors had to memorise. 'Realise that the obligation [of loyalty to the emperor] is heavier than the mountains, but death is lighter than a feather,' one section read. In other words, every soldier had to be prepared to sacrifice himself for the 'divine father'.

33. The *Sakura* Ideology

From the 1880s onwards, Japanese nationalists often cited Norinaga Motoori's poem linking the cherry blossom to the 'Yamato spirit' as a way to connect the *bushidō* philosophy with the symbolism of the nation's iconic flower. And when Nitobe wrote his best-selling *Bushidō* book, he noted how the cherry was 'the emblem of the Japanese character' and was 'ever ready to depart life at the call of nature'.

In the late nineteenth century the Meiji government started to incorporate images of cherry blossoms into the military, as well as in educational and cultural events, reinforcing the idea that a soldier should be prepared to die for the emperor like the cherry blossom. After 1870, for example, the Imperial Japanese Navy's badge became a combination of an anchor and a cherry blossom, featuring in the insignia worn on caps, shirt collars, sleeves, shoulders and buttons. The Imperial Japanese Army also incorporated a cherry in their designs.

The *Forty-Seven Rōnin*, the well-known drama about the former samurai who avenged the forced suicide of their master, also played a key role in linking the cherry with death. In the 1748 staging of the play, known in Japan as *Chūshingura*, loyalty to one's master had not been the main theme. Instead the *rōnins'* main purpose in that performance was to rebel against the shōgun for treating their master, Naganori Asano, unfairly. By the late nineteenth century, however, loyalty was increasingly being emphasised and it had become customary for cherry blossoms to be used as dramatic symbols throughout the play. In these newer iterations the play's culmination came as Asano committed *hara-kiri* (ritual suicide) at the height of the cherry-blossom season and as flower petals drifted onto his body.

In fact the *Forty-Seven Rōnin* became so seminal to Japanese culture that the proverb 'The cherry is first among flowers as the samurai is first among men' became popular, after it was mentioned in the play. The proverb originally meant what it said: in the same way that the cherry is the finest type of flower, so the samurai is the finest type of man. By the 1930s, however, the interpretation had been twisted: in the same way that the cherry blooms and then falls, the warrior who fights and then dies is the finest among men. For soldiers poised for battle, the analogy was abundantly clear.

One man, Kiyoshi Hiraizumi, had more influence than most in promoting these jingoistic messages. Then Professor of Japanese History at Tokyo Imperial University, Hiraizumi gave 'spiritual' lectures in the 1930s to the military, police, educators and students about the 'Japanese spirit', in which he linked his belief in Shinto practices with cherry blossoms, *bushidō* and the emperor, to fashion an ideology of Japanese supremacy. In a booklet entitled *Loyalty and Morality*, Hiraizumi was blunt:

> In case of emergency, we need to fall like cherry blossoms for the emperor. We Japanese have admired cherry blossoms since ancient times with this spirit. We don't rejoice over the blossoms, we rejoice over the flowers' falling and their heroic attitude. This is the basic spirit of Japan's appreciation of cherry blossoms.

Hiraizumi's comments were given additional weight by the fact that the blossoms of the *Somei-yoshino* cloned variety, which dominated the landscape by the 1930s, all bloomed and fell at the same time. That was a relatively new phenomenon. Before the mass plantings of *Somei-yoshino* in the late nineteenth and twentieth centuries, cherry blossoms in Japan did not give such an impression. Indeed, what few realised was that the *Somei-yoshino* cherry had not even existed before the 1860s.

34. The *Somei-yoshino* Invasion

T he *Somei-yoshino* cherry was undeniably beautiful. Even Collingwood Ingram, who sought to spread as many varieties as possible, said that 'a well-grown tree with its still bare boughs covered in soft pink bloom presents a very lovely sight'. But where had this tree come from? Even today the origins of the *Somei-yoshino* variety are unclear. All that is known is that the trees started being sold sometime during the 1860s in Somei village, a rural community in north Tokyo that boasted an unusually large number of nurseries during the Edo period. At the time it was unclear which the variety's parents were, although today most experts agree that *Somei-yoshino* is a hybrid of the Ōshima and *Edo-higan* wild-cherry species.

The tree had several convenient attributes. Whereas *Yama-zakura* cherries – the popular wild cherry – took a decade or more to reach maturity, *Somei-yoshino* trees grew to full size in about five years in the right conditions. The tree was also economical to grow, thanks to its high success rate when grafted. All this meant that only two decades after the Meiji Restoration, *Somei-yoshino* trees had replaced wild *Yama-zakura* in key Tokyo locations that had been renowned during the Edo period for their individual diversity.

Throughout Japan the government also planted the clones in castle grounds, to make the link between the flowers and warriors. In 1881 *Somei-yoshino* trees were even planted in front of the British Embassy (then known as the British Legation), by the UK's Envoy Extraordinary, Ernest Satow. They also graced the banks of the north-west moat of the Imperial Palace.

Among the famous cherry-viewing sites created in the capital during the Edo period, only the embankment at Koganei remained untouched by this invasion. There, thousands of *Yama-zakura* had been planted in the eighteenth century by the shōgun for the general population to enjoy. Since these trees were healthy, the government could not justify replacing them. But by the beginning of the twentieth century more than one-third of the cherry trees in prominent places in Tokyo were the lookalike *Somei-yoshino* variety.

That was just the beginning. Successive city and provincial governments around Japan celebrated major events by planting *Somei-yoshino* trees. Victories in the Sino-Japanese and Russo-Japanese Wars, the enthronements of Emperors Taishō and Shōwa, plus multiple other Imperial Household events all prompted plantings. So, too, did the establishment of public parks and the construction of schools and other buildings.

After the devastation of the Great Kantō Earthquake in 1923, tens of thousands more trees were planted along riverbanks and roads, and in parks, to give much-needed colour and vitality to devastated parts of Tokyo and Yokohama. It wasn't long before the widely planted wild cherries, such as *Yama-zakura* and *Edo-higan*, were driven back to the mountains from which they had come.

And predictably, when the Japanese government sent cherry trees overseas as goodwill gifts, *Somei-yoshino* dominated. About 60 per cent of the cherry saplings sent to Washington in 1912 were *Somei-yoshino*. The rest were a mixture of ten cultivated varieties.

The planting of the clones altered the cherry-blossom experience for *hanami* lovers and aficionados. Until the Meiji period, wild cherries and many cultivated varieties grew in the gardens of *daimyō* and in public spaces. The leaves and flowers of each of these cherries differed in colour, size and shape and they blossomed at different times. Now, in areas where *Somei-yoshino* dominated, the flowers blossomed simultaneously. This was a spectacular sight, yet as Ingram pointed out, it lacked variety. Ingram loved Koganei's *Yama-zakura* avenue precisely because the trees there were wild and distinct from one another. By the late 1920s this had become a rare vista.

'The famous old cherry tree locations used to have ancient *Yama-zakura* and some *Sato-zakura* [cultivated or man-made cherries],' Kiyoshi Inoshita, the director of Tokyo's Public Parks and executive secretary of the Cherry Association, noted in 1936. 'But these were replaced by *Somei-yoshino* before anyone noticed. When planting new trees, there's no hesitation in choosing only *Somei-yoshino*. Consequently, the vast majority of cherry trees have ended up being this one variety.'

Connoisseurs such as Inoshita, Seisaku Funatsu and Manabu Miyoshi mourned the loss of diversity that accompanied *Somei-yoshino*'s advance. Even during the normally placid meetings of the Cherry Association, criticism against these cherries was heated, according to Inoshita. 'No flower has ever been so disparaged, lambasted or repudiated so caustically by experts and cherry lovers alike as *Somei-yoshino*,' Inoshita wrote. 'While *Somei-yoshino* ought rightly to be hounded out of the cherry tree world bloodied and bruised, in reality it continues its advance at its leisure, unaffected by the protests. It's as if *Somei-yoshino* is asserting that it is the only legitimate cherry.'

Masuhiko Kayama, the teacher and cherry enthusiast who had helped Ingram and Tōemon Sano with *Taihaku*'s return to Japan, was equally dismissive. He called for *Somei-yoshino* trees that had been planted at two landmarks – Kyoto's Ninna-ji Temple and Kamigamo Shrine – to be replaced by either *Yama-zakura* or another 'more noble variety'. In spite of all this criticism, *Somei-yoshino* ruled supreme and, to ordinary Japanese people, this sole variety became the nation's one 'true' cherry.

35. 100 Million People, One Spirit

By the mid-1930s the association between cherries and national identity was firmly established in Japan. And Japanese diplomats began to export the imagery internationally, using the cherry blossoms as peaceful propaganda to mask the nation's less peaceful intentions. As Japanese–American relations deteriorated, for example, so the Japanese ambassador in America, Hiroshi Saitō, who had been Roosevelt's friend since meeting him in Washington in 1911, used the analogy of the cherry blossom and the rose to try and find common ground between the two nations. In a radio speech broadcast in the US at the height of the cherry-blossom viewing in April 1934, Saitō explained:

> There is earnestness in the rose, but animation in the cherry. The rose holds to life till the very last, while the cherry makes light of death and dances down in the breeze.
>
> The rose with its thorns stands for rights, while the cherry for duty with its unobtrusive colour. The rose

is individualistic and self-assertive. The cherry is to be enjoyed in clusters, each flower losing its individual identity in the making of the whole.

But the cherry and the rose have one thing in common: beauty, which is a joy forever.

My friends, the East has presented us her cherry blossoms, and the West her rose. In our search after differences, we often forget that the Earth is round and the so-called farthest East touches the farthest West.

Ambassador Saitō's flower comparisons echoed many of the thoughts of Inazō Nitobe, the author of *Bushidō*, three decades earlier. But the American public remained unconvinced about Japan's goals. Saitō was a popular diplomat, and his seven-year-old daughter, Sakiko, was named Queen of the Cherry Blossoms at the 1937 Cherry Blossom Festival in Washington. Yet he had the unenviable task of trying to justify Japan's aggression in China in the 1930s, as well as the country's increasing isolationism.

In Japan few people publicly questioned the new cherry ideology. Challenging the establishment had become a treasonable offence after the enactment of the Maintenance of the Public Order Act in 1925 and of subsequent laws that further tightened the government's control over free speech. Anyone who challenged the emperor system or the Japanese Empire was severely punished or sentenced to death.

This posed a problem for some members of the elite Cherry Association. Although most members were firmly of the establishment, there were still differences of opinion.

One dissenter was Yoshio Yamada, a longtime contributor to the *Sakura* journal and a Japanese classics scholar. He was also a realist, who in 1938 argued in an influential magazine, *Chūōkōron*, that there was no evidence to link cherry trees and the *bushidō* spirit. 'Falling petals are a phenomenon in many plants, and cannot therefore be seen as characteristic of cherry blossoms,' he wrote. It was a theme to which he returned three years later in the *Asahi* newspaper, one of Japan's most popular publications. There, Yamada wrote about the interpretations of Norinaga Motoori's poem, written in 1791, linking cherries with the Japanese spirit:

> One can argue that the cherry blossom is the national flower of Japan. But if the argument is that the cherry blossom is in any way a symbol of the Japanese spirit, I should be hard pressed to agree.
>
> I cannot agree that those who defend this position in terms such as 'the cherry tree's blossoms fall gracefully' or those who say that it aligns with *bushidō* appreciate the cherry blossom's true nature.

Rather, Yamada said, the correct way to appreciate cherries was simply to love their beauty.

It is not known how Yamada's colleagues greeted his arguments, and it is surprising they were published in the first place. For the minority of members who resisted the proliferation of *Somei-yoshino*, their voices were barely heard at a time when dissent was subversive and diversity had become anti-Japanese. Instead the Cherry Association, like other societies and national institutions, was swept along on the militaristic current. The 1940 issue of *Sakura* noted that association members had donated cherry-tree *bonsai* to the Anti-Comintern Pact countries and to the conquered Chinese province of Manchuria, where tens of thousands of Japanese had moved to work in coal, iron and gold mines and heavy-industry factories.

After the war against the US and Britain began in December 1941, the journal's writers joined in the euphoria of Japan's victorious invasions of Hong Kong, Malaya, the Philippines, Indonesia and Burma. The 'new' Japanese Empire now had an abundance of oil, rubber, tin, rice, soya and other materials with which to fight the US and Britain.

The government had announced in 1937 that the population of the Japanese Empire, including Korea, Taiwan and other territories, had topped 100 million. That prompted Japanese newspapers and radio stations to coin two terms – 'One hundred million people, one spirit' (*Ichioku isshin*) and 'One hundred million fireballs' (*Ichioku hi no tama*) – to refer to the unified morale of the Japanese people.

In April 1942 the Cherry Association's annual meeting opened to the sound of air-raid sirens and ended prematurely. Earlier that month the US had made its first air strike against the Japanese mainland, in an operation now known as the Doolittle Raid. That year's *Sakura* journal was unabashedly militaristic in tone, and its vitriolic opening comments were written by Kiyoshi Inoshita, the Tokyo parks director, under the pen name 'Flower Protector':

At the foundation [for victories of the Imperial Japanese Army], there lies an unspoken abundance of the time-immemorial cherry blossom spirit. Within this, our soldiers earnestly wish to die bravely for our honoured emperor.

The noble figure of the cherry tree, bright and fragrant in the morning sun, is the spirit of one hundred million Japanese citizens. After they blossom, it is the Japanese character to fulfil their duty bravely.

Win, win, win gloriously.
One hundred million people, one spirit.
In a fireball!!

In the same issue the association's chairman, Duke Nobusuke Takatsukasa, the aristocrat whom Collingwood Ingram had befriended in England in 1924, wrote a similarly patriotic essay entitled 'March Forth, Cherry Blossoms':

Cherry blossoms are considered to be indispensable for the Japanese empire. At a time when the blossoms cover the mountains and fields with dazzling brilliance and when the war is achieving every possible victory, we are struck with deep and powerful emotions when we think about cherry blossoms.

With the loss of some of its key early members, it was all too clear by 1942 that the Cherry Association had lost its rudder. Seisaku Funatsu, the cherry doyen and protector of the diverse Arakawa River varieties, had died in 1929. A decade later Manabu Miyoshi, the cherry professor from Tokyo, passed away too.

In 1943, amid the madness of the 'One hundred million fireballs' and as the tide turned against Japan in the war, the Cherry Association released its final journal and slipped quietly out of existence. Yet the cherry blossoms' 'role' in the Pacific War was by no means over. Since the US victory at the Battle of

Midway in June 1942, Japan had been on the back foot, steadily retreating across the Pacific Ocean from lands it had occupied earlier in the war.

As defeat — the ultimate humiliation — was not an option for the military high command, the generals had decided on a last resort: suicide attacks on Allied ships by *kamikaze* pilots. To persuade the public and pilots alike that the fighting must continue, Japan's leaders increasingly invoked the symbolism of falling cherry blossoms, and the flower's links to the emperor and to the philosophy of *bushidō*.

36. The Cherry and the *Kamikaze*

On 20 October 1944 the 'inventor' of the *kamikaze* tactics, 53-year-old Admiral Takijirō Ōnishi, spoke to the pilots of twenty-four planes that were preparing to take off during the Battle of Leyte Gulf in the Philippines. The men all knew they were flying off to their deaths.

'Japan is in grave danger,' Ōnishi said. 'On behalf of your one hundred million countrymen, I ask this sacrifice of you and pray for your success.' As head of the First Air Fleet in the Philippines, he added: 'You are already gods, without earthly desires. But one thing you should know is that your crash dive is not in vain.'

Before he spoke, Admiral Ōnishi had presented the pilots with a poem that he had composed:

Today, in blossom
Tomorrow, scattered by the wind

Life is so like a delicate flower
How can one expect its fragrance to last for ever?

The *kamikaze* Special Attack Forces, known in Japan as *tokubetsu kōgeki tai* (or *tokkō-tai* for short), were grouped into the *Shikishima, Yamato, Asahi* and *Yama-zakura* units. These names were all references to Norinaga Motoori's poems, deliberately linking cherries with the Japanese spirit. In the first wave of attacks the planes dive-bombed into a US aircraft carrier and other military vessels in the Leyte Gulf.

Pink cherry blossoms on white backgrounds were painted on the fuselage of many *kamikaze* planes, and their bombs were sometimes called 'cherry bombs'. The attacks themselves were regarded as acts of *sange*, a ritual of scattering petals in a Buddhist memorial ceremony. All these concepts were used to glorify and beautify the pilots' deaths, the idea being that the men experienced a divine death as a flower – a cherry blossom. After death, they became *gunshin*, or military gods, and were reborn as cherry blossoms at the Yasukuni Shrine in Tokyo.

Six months after the first wave of attacks, the military leaders ordered further *kamikaze* strikes on US ships and the British Pacific Fleet, which were patrolling the seas off Okinawa. After American troops had landed on the Kerama Islands in Okinawa on 26 March 1945, Japan's top officials thought it was only a matter of time before they invaded the southernmost island of Kyushu. Pilots were recruited from all over Japan. Most were aged between seventeen and twenty-five, including many university students. Early in the war students had been exempt from conscription, but had lost this privilege in October 1943 as the battle losses mounted and were now being targeted as *kamikaze* pilots.

Many of these pilots gathered at a feudal town called Chiran, 12 miles from Kyushu's southern coastline, where there was an air base. Makeshift A-frame wooden huts had been hurriedly built in the woods near the base, which is today surrounded by plantations growing green tea. The huts were hidden amid cedar trees to evade US raids, and the pilots spent a few nights there, before taking off for their final flight.

On 6 April 1945, a 20-year-old Imperial Japanese Army pilot, Second Lieutenant Kazuki Kamitsu, wrote a poem there that encapsulated the essence of the cherry ideology:

For the glory of the emperor
What is there to regret?
As a young cherry
Life is most worthy when falling.

I had come across this poem while researching Collingwood Ingram's passion for cherry blossoms. It struck me how closely the musings of individuals mirrored those of the government. I was eager to know whether other *kamikaze* pilots had talked or written about *sakura* before they flew off to die. So on one of my visits to Japan I jumped on a *shinkansen* bullet train in Tokyo, transferred to ever-smaller trains in Kyushu and eventually arrived by local bus in the tiny town of Chiran in Kagoshima Prefecture.

Chiran's major attraction these days is its grandiose Peace Museum, a government-funded institution that attracts coach-loads of schoolchildren and elderly Japanese visitors. In Chiran, I was surprised to find how much the image of cherry blossoms dominated the final days of many *kamikaze* pilots. Two of the Imperial Japanese Army captains were called Toshio Anazawa and Seiichi Kishi, and the last night of their short lives was Wednesday, 11 April 1945. By chance, it also happened to be the final day in the life of the longest-serving US president, Franklin Delano Roosevelt.

37. Falling Blossoms

FDR, as the thirty-second US president was known, spent his last evening in a spartan bungalow surrounded by pine trees at a hot-springs resort in Georgia. Roosevelt, who had been president since 1933, had left Washington D.C. as the last *Somei-yoshino* cherry blossoms of spring – a gift from Japan in 1912 – fluttered down onto the Potomac soil. At the so-called 'little White House' in Warm Springs, Roosevelt was hoping to recover from a series of illnesses and wartime crises that had visibly weakened the nation's 63-year-old commander-in-chief.

Far away in the Pacific, tens of thousands of US Navy personnel were massed on ships near the tropical Japanese islands of Okinawa, preparing for invasion. Aboard the USS *Whitehurst*, a destroyer escort, two young men – Irving Paul from Tucson, Arizona, and Odell 'True' Lofton, from the rural hamlet of Dennard, Arkansas – were on radar-picket duty. Together with the 211 other battle-weary sailors aboard, they were hoping to detect Japanese aircraft before they could bomb the long convoy of supply ships.

About 650 miles north of the USS *Whitehurst*, near the Chiran Air Base, Captain Toshio Anazawa lay on a straw-filled futon covered with a thin blanket and thought about his fiancée, Chieko Sonoda. 'The cherry blossoms have already fallen here,' the 23-year-old Chūō University graduate wrote to her. 'The young green leaves I love so much will soon usher in a new spring.' Close by, in another A-frame wooden hut, 26-year-old Captain Seiichi Kishi, wrote a poem to his parents:

The cherry blossoms are falling
One after another
I also want to fall now
Leaving the scent
In Yamato country.

Anazawa and Kishi knew they faced death the following day, 12 April 1945. They had no choice, for their fate had been sealed months earlier. They weren't dying just for Japan. They were dying for the nation's 124th emperor – dying to protect an

Kamikaze *plane with cherry blossom*

imperial system that allegedly stretched back to Emperor Jinmu in 660 BC.

That evening Anazawa, Kishi and their *kamikaze* unit held a farewell party, downing bottles of *sake* and singing army songs at the top of their voices. The most popular was 'Cherry Blossom Brothers' (*Dōki no Sakura*) – the same song, but a different verse, that my father had sung to me from his bed in a care home seventy-one years later:

You and I are cherry blossom brothers
Blooming together in the military academy garden
Having blossomed, we must scatter
Let us fall magnificently for our country.

When the pilots woke at about 4 a.m. they dressed in their combat fatigues and were driven to the air base. There they downed a ceremonial cup of *shōchū*, a strong distilled drink, screamed the battle cry '*Tennō Heika Banzai!*' ('Long Live the Emperor!') and climbed into the cockpits of their bomb-laden Nakajima-Ki 43 aircraft.

Around his neck Anazawa wrapped a white scarf, a gift from Chieko. The letter that he had completed to her the previous night was calm and dispassionate: 'I, Anazawa, no longer exist in the world of reality. From now on, you have to live in the reality of each moment. Don't waste your time with trifles from the past. Find new ways to be happy in the future.'

As his plane taxied down one of the base's two runways, Shōko Maeda, a 15-year-old pupil from Chiran Girls' High School, was watching the departures with her fellow students. She wrote in her diary:

Anazawa's plane passed in front of me. We all waved cherry blossom branches to say farewell. Then Anazawa, who was wearing a headband, smiled and gave a salute over and over. CLICK! I looked behind me at the sound, and a cameraman had taken a photograph of us. I stood there for a long time, gazing at the southern skies. My eyes filled with tears.

That photograph has become iconic. Taken by a photographer standing behind Maeda, it shows the so-called *Nadeshiko* girls (named after a pink carnation) grouped alongside the runway, waving cherry-blossom branches as the pilots took off. Published in the Osaka edition of the *Mainichi* newspaper on Tuesday, 17 April 1945, the photo was banned after the war, but was secretly preserved by the newspaper until it was made public again in the 1960s.

The cherry-blossom imagery, of course, was pure propaganda, designed to rally the population when the war in the Pacific

The Nadeshiko *girls wave farewell to a* kamikaze *pilot*

appeared to hang in the balance. In reality, Japan's 'cherry blossom' war was long since lost and was hastening towards its nuclear reckoning.

Thursday, 12 April 1945 was one of many deadly days in the war's final phase, but Roosevelt's death after a cerebral haemorrhage gave it added significance. That evening, Vice-President Harry S. Truman took the oath of office. By the end of the day Anazawa, Kishi and forty other *kamikaze* pilots were dead. They had either dive-bombed into US Navy ships or been shot down en route to their destinations. More than 200 American servicemen died on 12 April, on board at least seventeen US Navy vessels targeted by the Japanese planes. They included forty-two men aboard the USS *Whitehurst*.

'Tokyo Rose [a name often given to female English-speaking collaborators who broadcast to Americans] had warned us the night before that there would be an all-out air attack by the *kamikaze*,' James Nance, a Texan who was on board the USS *Whitehurst*, recalled. 'The suiciders came howling down with death in their hearts. About 2.30 p.m. I was writing a letter to my wife Kathryn when the general alarm sounded and we all dashed to our battle stations. We spotted five dive bombers approaching from the west.'

One was carrying a 500-pound, delayed-action bomb that exploded about 50 feet from the side of the ship with such force that it killed or seriously injured more than thirty crew members. A second plane crashed into the USS *Whitehurst*, sending flaming petrol into the Radio Room and into adjacent passageways. The explosion killed Irving Paul outright, when the door of the Radio Room slammed into him after being blown off its hinges, and 'True' Lofton died of asphyxiation as he tried to escape the fire and smoke.

In all, between October 1944 and August 1945, about 3,800 Japanese pilots died in *kamikaze* attacks, including 1,036 army pilots who took off from the Chiran Air Base. More than 7,000 Allied personnel were killed.

38. Tome's Story

S ince 1975 many of the *kamikaze* pilots' poems and testi- monies, mostly written to their parents, have been exhibited at the Chiran Peace Museum. Other documents are in libraries or in private hands.

The museum also contains videos of the *kamikaze* attacks. There is a replica of the A-frame hut where the pilots spent their final nights, and photographs of the pilots eating together, arm-wrestling, meeting local people and singing songs in small groups. Their smiles are radiant. They seem joyful at the pros- pect of imminent death. But were they?

The personal testaments were touching in their naivety, but they were unconvincing to me. Did these young men genuinely believe they were falling for the emperor? Did they really die with smiles on their faces?

The official line was that the men had volunteered willingly. 'They did not hesitate to become *kamikaze* pilots,' the museum's official booklet said, of a photo of five smiling 'boy pilots' cud- dling a puppy. Three were seventeen years old and the other two were eighteen.

When I returned to the nearby inn where I was staying, I told the manager's wife why I was in Chiran. 'You should meet a man I know,' she said. 'His grandma helped look after the boys as though she was their own mother.' She arranged a meeting that evening.

And so it was that I met Akihisa Torihama, who turned out to be the true guardian of the Chiran *kamikaze* pilots. He was the person I had been searching for, to explain my apprehensions about the museum. In a restaurant that Torihama owns, he sat me down on a wooden bench near the entrance and began to talk in a soft but determined voice.

His grandmother, Tome Torihama, had owned a restaurant called *Tomiya Shokudō* on the main street of Chiran. The military had designated it a place where the *kamikaze* pilots could go to eat on their days off. Although Tome died in 1992, she is still the most renowned inhabitant of this former samurai settlement. She became the mother of the *kamikaze*, their confidante, their friend and a shoulder to cry on.

'They were all young boys, away from home,' Akihisa Torihama said. 'They needed love and care. Tome welcomed them, comforted them and fed them her home-made meals without charge. Some of the boys had lived in Chiran a long time before they became *kamikaze* pilots, because there was a school there to train young airmen. So Tome knew many of them already and she went out of her way to look after them. She was forty-three in 1945 and they all called her *okā-san* [mother].'

'Grandma was always telling me that the boys were lonely, in great agony and afraid to die,' he said. 'The military training was really harsh. The boys came to the restaurant dragging their feet, or with swollen eyes or mouths because they had been beaten up. None of them went with a smile.'

Tome, he said, encouraged the boys to write private letters to their parents, because the military censored their official testimonies and letters. 'She often got up at dawn and went to the next village to post the letters secretly. It was very dangerous. She risked her life. After the boys took off, she wrote to their parents to tell them about their final days.'

The following day Torihama showed me around the small private museum he has opened on the original site of his grandmother's restaurant on Chiran's main street. Demolished after the war, the wooden two-storey eatery has now been rebuilt as a replica. On one wall is that seminal photograph of the high-school girls waving cherry branches. One of the girls was Tome's daughter, Akihisa Torihama's aunt. According to Torihama, the military had asked the girls to collect cherry-blossom branches that morning and to stand next to the runway. 'That was probably the only time the girls were allowed inside the air base,' he said. 'Do you think civilians would have been allowed on the runway? Never.'

Torihama's museum exhibited many documents and photos that indicated the true state of the young men in their final days. It was clear that while the pilots all pledged their allegiance to the emperor before they took off to die, beneath the surface they harboured doubts, anguish and sadness. Many of them were forced to accept the ideology and had no way of escaping their fate.

Perhaps the most striking example of the young *kamikaze* pilots' ambivalence displayed in the museum is that of a brilliant economics student from Keiō University, Ryōji Uehara, who died at the age of twenty-two. Uehara confided in Tome his fear that Japan would lose the war, Torihama said. These were dangerous comments, punishable by execution as a treasonable offence, if spoken in public. A glass case on the museum's ground floor houses Uehara's final testimony.

On 10 May 1945, the night before his own *kamikaze* mission, Uehara refused to write an 'official' final testimony when asked to do so by the military because he knew it would be censored. Instead he met a friendly military photographer, Toshirō Takagi, who asked him to record his true feelings. Uehara wrote his thoughts on a piece of paper and gave it to Takagi, who secretly sent the paper to the student's parents.

Uehara addressed his testament to the 'Japanese people' rather than to the state or the emperor. It began with his 'official' words: 'I am extremely honoured to have been chosen as a glorious *kamikaze* pilot for my beloved country.' And it continued:

Authoritarianism and totalitarianism may thrive for a little while but they are destined to be defeated. My wish

Ryōji Uehara

was that my beloved Japan would become a great country like the one-time British Empire. But that is now all in vain.

Kamikaze pilots are mere machines without any personality, emotion and reason. Their sole purpose is to attack the enemy's aircraft carrier like an iron particle. It is a suicidal act. It is only in Japan where this 'spirit' [*Yamato damashii*] prevails.

Another glass case contains a second testament, which Uehara had secretly written on a visit home. 'I admired liberalism because I thought that was the only way Japan could thrive for ever,' Uehara wrote. 'Victory will clearly go to the side where a more natural philosophy based on human nature prevails.'

How, I wondered, must this intelligent, passionate young man have felt at dawn on 11 May 1945, as he walked towards the Chiran runway with five other pilots? As the engineers checked their planes, the six pilots in Uehara's group turned to the east – the direction of the Imperial Palace in Tokyo – and bowed deeply. The men then made a circle and started singing, clapping their hands rhythmically:

> If you are a man
> Don't regret this fleeting life
> What's important is the flower's falling and man's courage
> Leave your luck to the wind
> If you are a man
> Just act and fall.

At 6.15 a.m. the planes took off one by one. At 9 a.m., according to a letter later sent to the family by the military authorities,

Uehara's plane crashed into an unnamed Allied ship. That day no fewer than 104 *kamikaze* planes had taken off from Chiran and other airports in Kyushu. No American ships were sunk, but the *kamikaze* attackers damaged an aircraft carrier, the USS *Bunker Hill*, and two destroyers.

I felt a profound sadness. Anger, too. About the cult-like worship of the emperor and the government that had used the cherry blossom as part of its perverted ideology. About the military regime that had sent so many young Japanese men to their deaths and caused the unnecessary loss of so many Allied forces.

The *kamikaze* attacks had little impact on the course of the Pacific War, and the American bombing of Japanese cities continued unabated. The war ended after the US dropped atomic bombs on Hiroshima and Nagasaki on 6 August and 9 August respectively. The bombs killed more than 129,000 people. Japan was in total ruins. On 15 August, Emperor Hirohito gave a radio address in which he announced the country's surrender to a shocked and silent nation.

Yet the cherry ideology lingered. Hours after the emperor's speech, the 55-year-old Vice-Admiral Matome Ugaki, head of the Fifth Air Fleet, who had been directing the attacks on US and British ships near Okinawa, climbed defiantly into a two-seat dive-bomber and flew off with thirteen other planes. They were never heard from again, although Ugaki's last radio message was recorded at 7.24 p.m.

'I am going to Okinawa, where our men lost their lives like cherry blossoms,' it read. 'I will destroy the conceited enemy in the true *bushidō* spirit, with firm conviction and faith in the eternity of imperial Japan. Long live the Emperor.'

PART SIX

—

DARK SHADOWS

39. Children at War

In the early evening of Monday, 8 December 1941, Collingwood Ingram turned on the wireless in the sitting room of The Grange in Benenden and listened with Florence as Prime Minister Winston Churchill addressed the nation. A day earlier Japan had attacked and sunk American ships at Pearl Harbor in the US territory of Hawaii. Japan had also attempted to land troops in Malaya and had bombed both Singapore and Hong Kong. The US and Britain were now at war with Japan.

'When we think of the insane ambitions and insatiable appetites which have caused this vast and melancholy extension of the war, we can only feel that Hitler's madness has infected the Japanese mind and that the root of the evil and its branch must be extirpated together,' Churchill told the nation in an emotional broadcast. 'In the past, we had a light which flickered. In the present, we have a light which flames. In the future, there will be a light which will shine, calm and resplendent, over all the land and all the sea.'

In Washington D.C. on that same day, President Franklin D. Roosevelt told the largest audience in radio history that 7 December 1941 was 'a date which will live in infamy'.

As Churchill's broadcast ended, the Ingrams' thoughts inevitably turned to their four children and five grandchildren. The Ingrams' only daughter, Certhia, lived nearby with her farmer husband, Gerald, and their four-month-old daughter, Veryan. But all three of the Ingrams' sons were in the armed forces. War with Japan would clearly mean more fatalities and casualties for Allied troops, and greater anguish for their families. The Ingrams' eldest son, Ivor, had so far been lucky. Having volunteered to join the Royal Air Force, he was based in Britain in December 1941. Happily married to Winifred, they had two children, three-year-old Jennifer and baby John, born earlier that year. Meanwhile, Mervyn, the Ingrams' second son and a doctor, had been dispatched to North Africa as a captain in the Royal Army Medical Corps, leaving his wife, Joan, in England with four-year-old Jane and two-year-old Collingwood. Alastair, the Ingrams' youngest son, had become an army officer with the Royal Artillery, after leaving Winchester College. Having been sent to Hong Kong in early 1940, he had met there, and fallen in love with, a British military nurse named Daphne Van Wart.

The Ingrams had not met their son's sweetheart, but they sensed how smitten Alastair was with Daphne, who had arrived in the British colony of Hong Kong from Liverpool in September 1940. Born in India, Daphne had never known her mother, Annie, who died two weeks after her birth; her father, Reginald, who was from a Flemish Huguenot family, was headmaster of India's elite Chopasni School in Jodhpur, to which many colonial families sent their children.

Unperturbed by Japan's expansionism in China and Manchuria, Hong Kong had remained a lively party town for the British troops and nurses through the early years of the war. On days off, and after her shifts at the Bowen Road Military Hospital, Daphne would meet Alastair to go dancing and sailing. Within months, the pair had started planning a life together.

Yet the good times were not to last. First, Alastair was posted back to Europe in 1941, leaving Daphne alone in Hong Kong. Then on 8 December, just four hours after Japan attacked Pearl Harbor, the Japanese army bombed the British air base in the Kowloon peninsula and invaded from China by land. The Battle of Hong Kong had begun. Caught off guard and ill prepared, the Allied forces were outnumbered, outgunned and outmanoeuvred. They could do little to prevent parts of Hong Kong being occupied and ransacked by the Imperial Japanese Army.

40. Black Christmas

Daphne's fifteen months of colonial-style living were over. After the initial Japanese invasion in December 1941, she was transferred to a 400-bed field hospital that had been set up at St Albert's Convent in the centre of Hong Kong Island. Day by day the number of casualties brought into the hospital with life-threatening injuries increased, as the Allies fought desperately to repel the Japanese.

On 18 December, Japanese forces began bombarding the hospital grounds, hitting the nurses' canteen at lunchtime. Brenda Morgan, a colleague of Daphne, was killed instantly and the nurses' matron, Kathleen Thomson, was injured in the blast. Daphne herself escaped because she was taking shelter in the cellar. On 23 December the Japanese army captured the hospital and tied up most of the medical staff.

Much worse was to come at another hospital on Hong Kong Island, where Daphne knew many of the nurses. That hospital had been set up at St Stephen's College, located in the south of the island. There, early on Christmas Day, between 150 and 200 Japanese soldiers stormed into the building.

More than fifty wounded British soldiers were slaughtered as they lay in their beds. Then, after locking the female nurses in a room upstairs, the Japanese soldiers raped and killed five Chinese and three British nurses. Four more British nurses were gang-raped that day, but survived, including a Scottish woman called Molly Gordon, who was to become one of Daphne's closest friends. At the time of the massacre all the women were wearing nurses' uniforms and Red Cross armbands.

The remaining male patients and staff were herded into a small room from which, at frequent intervals throughout the day, one or two at a time were taken into a corridor and shot. On the afternoon of 25 December the governor of Hong Kong, Sir Mark Aitchison Young, formally surrendered in person to the Japanese. The events of that day became known in Hong Kong as 'Black Christmas'.

It was an agonising time for Alastair and the Ingrams. With little information about the fate of the soldiers, nurses and other foreign residents in Hong Kong, no one even knew whether Daphne was alive. It wasn't until late 1942 that the nurses could write postcards home to let their families know they were safe. Daphne, it turned out, had been transferred with other medical staff to yet another hospital, called St Theresa's, in Kowloon in February 1942, where she was 'appalled at the state of the patients, emaciated and desperately ill and suffering from starvation'.

Sick Allied troops were regularly brought to the hospital from a prison camp set up by the Japanese, mostly suffering from the highly contagious disease of diphtheria. The nurses did not have the antitoxin required to treat the disease and were given virtually no medicine by their captors.

After six months at St Theresa's hospital, in August 1942 Daphne was moved to the Stanley Internment Camp, which the Japanese had created at St Stephen's College, the site of the Black Christmas massacres. There she and some fellow nurses lived for the following three years as civilian prisoners of war, amid 2,800 other, mostly British, American, Norwegian and Dutch POWs. Malnutrition was the norm, and sicknesses such as diarrhoea, diphtheria and beriberi were endemic. During the forty-four months that the camp was open, 121 people died, including seven who were executed for possessing a radio. Yet the Stanley camp was not as inhumane as most other camps run by the Japanese military in the Far East. The POWs could walk around the site freely, as long as they bowed deeply to their guards and obeyed all the rules.

One POW, Dominica Lancombe, was just six months old when she was placed in the Stanley camp. For three and a half years she slept in a chest of drawers next to her mother, Elisabeth, who used the concrete floor as a bed. Her earliest memories – recollections of the Stanley camp – echoed Daphne's. 'My main memory is fear, fear of everything,' said Dominica, who was seventy-six years old when we talked at her London home, a couple of months after I met her at a gathering of former POWs and their relatives. 'I had fear of being late for roll call every morning. Fear of being beaten if I didn't bow to every Japanese soldier. Fear of not eating. I was always hungry, even though I was only an infant. Mum queued for everything – a cup of hot water in the morning and one in the afternoon, a ball of rice at 11 a.m. and again at 4 p.m. To get more food, people would barter over the fence with the Chinese, so we occasionally had fruit or vegetables or an egg.'

In addition to these hardships, Daphne was appalled at the way the Japanese captors treated the female nurses. In the British army, nurses enjoyed high social standing. In Hong Kong the Japanese captors pushed them around and made them do menial chores. And while male POWs received a little pay for their labour, the nurses received nothing. For a proud Englishwoman, her second-class status was just one more reason why Daphne came to detest the country that had raised such men.

41. Protecting Benenden

In the three-and-a-half years that Daphne Van Wart was trapped in Hong Kong at the mercy of the Japanese, Collingwood Ingram was preoccupied with doing whatever he could to defeat the Germans. He completely supported Churchill's sentiment that Britain should resist the Nazis to the bitter end, whatever the cost.

As the commander of Benenden's Home Guard, Ingram organised a crew of ageing gardeners, butchers and other local volunteers, who worked in shifts every night from 9 p.m. to 5 a.m. patrolling the village to spot anything suspicious. While the threat of a German invasion of Britain had receded by 1941, Ingram's Home Guard unit was always on the lookout for German spies, escaped prisoners of war or shot-down pilots who might have parachuted into Kent. Regardless of the situation, Ingram was determined that his men could handle it.

'Surprise attacks and guerrilla tactics are of utmost importance,' he told the unit one night, adding that the best delaying tactic was to place obstacles on the ground that looked like mines. 'This sounds rather like a schoolboy trick, but I think it will work. An old motor car accumulator painted black, with a spark plug sticking into it and a wire running to both sides of the road, would look devilish suspicious – or a small square tin with some strange-looking contraptions attached.'

For almost five years Benenden existed in a state of emergency. In June 1940 the village's renowned girls' boarding school relocated its pupils to the seaside resort of Newquay in south-west England, to avoid the aerial bombing. The empty building in the village was transformed into a military hospital, to which thousands of wounded soldiers were sent. Meanwhile, many single women in Benenden joined the Women's Land Army, driving tractors, ploughing fields and picking apples, pears and soft fruits around the village to supplement the meagre food supplies.

Ingram would later recount stories about the Battle of Britain skirmishes between British and German aircraft that had taken place in the summer of 1940 above his home. Several planes crashed in the parish, with one damaging the front gates of The Grange. In a letter to relatives in Australia, Ingram said that he had had 'the pleasure of taking a German parachutist prisoner. [It was] not a very exciting experience, as the poor wretch was in a blue funk and was only too willing to give himself up.' The German pilot had become a British prisoner of war. Three other German pilots had been shot down and had died in Benenden and were buried in the church's graveyard.

Ingram relinquished his role as commander of the Home Guard in late 1941 or early 1942. He subsequently turned more of his attention to his cherry-tree research in the attic of The Grange, the quiet of the village now disturbed only by the drone of Allied aircraft flying over Benenden en route to bomb German cities. By 1944 it was clear from the newspaper headlines and BBC broadcasts that the Axis forces of Germany, Italy and Japan were in retreat. But the news did little to relieve the Ingrams' anxiety about Alastair, who was active as a Royal Artillery major in the Battle of Monte Cassino in Italy, or about his girlfriend, Daphne, marooned in the Hong Kong POW camp.

In the meantime Benenden faced a new threat in the summer of 1944. It came as the weakening Nazi regime began to bombard England with a huge new flying missile, the 27-foot V-1, which became known as the 'Flying Bomb' or 'Doodlebug'. Fired from launch sites in northern France and the Netherlands, the V-1s were aimed at London, but often fell prematurely along their flight path when their rudimentary autopilot system failed. More than 1,300 of them landed in Kent, including thirty-two that crashed in and around Benenden, killing five villagers and scores of animals. 'Often they would hit the farms and end up killing cattle,' recalled Patricia Thoburn, who was seventeen at the time and whose family frequently visited the Ingrams at The Grange. 'My parents' fields were also damaged, killing lots of cows and horses.'

Despite the threat from the V-1s, Germany's defeat began to look inevitable by the spring of 1945 as the Allied forces started to advance on that country. On 30 April, Hitler committed suicide. A week later, on 7 May, Germany surrendered. With that, the war in Europe was over and it was time for the Allies to celebrate.

42. Ornamental Cherries

'No more war!'
'No more blackouts!'

From Trafalgar Square in London to Times Square in New York to Red Square in Moscow, millions thronged their cities' main gathering places on 8 and 9 May 1945 to celebrate victory in Europe. In every village and town throughout the UK, citizens sang, drank and danced late into the night.

The Ingram family breathed a collective sigh of relief. The war had left Collingwood and Florence's offspring and their seven grandchildren unscathed. During the war Certhia had given birth to a son, Geoffrey, in 1943, and Mervyn's wife, Joan, had given birth to her third child, Charlotte, in 1944.

Benenden was safe again, no longer witness to overhead battles, the threat of invasion and frequent bombings. Aside from minor damage to some trees, Ingram's spectacular garden was intact. On 8 May, or VE-Day, as the day of victory in Europe became known, some late-blooming cherries, including the chrysanthemum-like *Kiku-zakura* and the white 'elephant' blossom *Fugenzō*, were still in bloom as Ingram inspected his collection.

Yet for all their joy at the end of the European conflict, the Ingrams knew that Alastair's girlfriend, Daphne, remained a prisoner in Hong Kong and it was unclear how that situation would be resolved. Towards the end of the war an American naval bombing raid had accidentally targeted the Stanley Internment Camp, killing fourteen people. Although that was tragic, the raid was evidence that the Allied forces were beginning to assert control. Food and water became ever more scarce and there was little or no electricity at the camp, but the POWs could sense that liberation was only a matter of time.

Finally, on 15 August 1945, Japan surrendered, and Daphne's POW life ended. She arrived back in England in November, after an arduous trip via Vancouver from Manila on the USS *Admiral C.F. Hughes*, an American transport ship. Aboard were just twelve nurses and 4,000 troops, including hundreds of British Commonwealth prisoners of war, many of whom were both mentally and physically sick, after years of suffering. Daphne and Alastair were reunited and the couple were soon engaged. They married in January 1947 at St Paul's Church in London's West End.

Alastair and Daphne's wedding day

One wonders how much thought Collingwood gave during those war years to the state of his beloved cherries in Japan? He would have found nothing to cheer him. Most of the wild cherries in the Japanese mountains still remained, but a majority of their urban cousins had perished. The cherries had been bombed or burned to a crisp by American air raids; uprooted to clear space for crops; chopped down for firewood to provide heat for homes and for cooking. The flimsy blossoms that had fallen each spring disintegrated into the soil, as an army of unforgiving hatchets slashed at the trees' noble trunks. And while the cherries in England still bloomed as if to salute the Allied victory, the cherries in Japan vanished amid the ashes of defeat.

Yet the varieties that Ingram had collected or imported from Japan lived on at The Grange. Secluded from the war, most had grown quietly and safely over the past decade. Ingram, who turned sixty-five in October 1945, had continued his cherry research throughout the Second World War and was soon ready to publish all the observations and illustrations he had made over the previous quarter-century.

The result was a triumph: a 295-page monograph called *Ornamental Cherries*, which combined the papers that Ingram had published for the RHS in 1925, 1929 and 1945, plus additional notes. It immediately became the definitive English-language guide to cherry trees as soon as it was published in 1948. Seven decades later, Ingram's passion rings out from every page, and the book remains a horticultural classic and a bible for cherry-lovers and researchers everywhere.

The book was dedicated to 'all who have planted cherry trees, whatever their creed, caste or colour may be. Consciously or unconsciously, they have made this world a more beautiful and a pleasanter place to live in.' Ingram wrote for all readers. For beginners, he outlined how to plant and grow cherries most efficiently and effectively. For experts, he explained the history of cherry-tree taxonomy and the scientific names of species and varieties.

What made *Ornamental Cherries* particularly compelling was that each of the 129 cherries described in the book grew in Ingram's own garden. He had collected or created them all. He described sixty-nine wild-cherry species, including ten from Japan, and others that grew naturally in China, Nepal, England, India, Tibet and elsewhere. He named one *Veryan's* cherry, after his seven-year-old granddaughter, Veryan Harden. Ingram also described sixty cultivated cherry varieties from Japan, dividing them into four groups by their colour: white, light pink, dark pink and greenish-yellow.

He had observed each tree's growth day by day, in the same way that his Japanese mentor, Seisaku Funatsu, had meticulously chronicled the cherries that grew on the Arakawa riverbanks. To Ingram, each blossoming tree had a distinct personality. The trees were his children and grandchildren, and he cared about them like the doting grandfather he had become.

Ornamental Cherries reads in parts like a tour around The Grange. One can almost smell the *Jō-nioi* cherry, as Ingram described its fragrance in the same way that a sommelier might portray a vintage wine. 'It permeates the air with a gorse-like scent, vaguely suggestive of crushed almonds,' he wrote of the tree's 'delightfully fragrant' snow-white petals. He was dismissive of the *Taki-nioi* variety's name, which in Japanese poetically suggested that the flowers possessed the fragrance of a waterfall.

Ingram had never noticed such a subtle resemblance, but he applauded the way in which the small white flowers showed themselves against the reddish-bronze tints of the unfolding young leaves.

Some cherries he lavished with praise. Of course *Taihaku*, which he had returned to Japan in the 1930s, stood 'supreme' and was 'by far the most beautiful of all the white cherries'. It was natural that the flowers he himself had named, including *Asano*, *Daikoku* and *Hokusai*, would be among his favourites.

Yet Ingram was a harsh critic of many other flowers in his collection. He disliked, for example, the prudish name of the *Pink Perfection* cherry, saying it reminded him 'of a spoilt Victorian child badly in need of a spanking'. The *Ōshōkun* cherry, named after a Chinese courtesan of breathtaking beauty, was 'unsatisfactory' because of its weak constitution, 'stunted and malformed branches' and 'gaunt, twisted boughs'. Meanwhile the disease-prone *Okiku-zakura* was 'both beautiful and ugly', like 'a beautifully dressed and heavily beringed woman of an uncertain age and unattractive figure'.

Poor *Okiku-zakura*! It wasn't easy being a cherry in the garden of such a demanding owner. But at least Ingram's cherries had survived, unlike most of their cousins in Japan.

43. Dark Shadows

For the inhabitants of The Grange, the 1950s were a time of stability and predictability. Life in the rural village of Benenden felt far removed from the difficult post-war life led by many people in British cities — and a world away from Japan's destitution. Despite his stature as an exceptional wartime leader, Winston Churchill and his Conservative Party lost the 1945 general election to the Labour Party, which nationalised key industries and put in place the free National Health Service.

The UK's colonies, meanwhile, started to disintegrate. India, Britain's biggest colony, was divided into India and Pakistan and both became independent in 1947. Burma split from the UK in 1947, and Ceylon (present-day Sri Lanka) in 1948.

Away from this geopolitical mayhem, Collingwood and Florence Ingram again enjoyed the sound of young voices at The Grange, three decades after they had first moved in.

One constant visitor in the early 1950s was their eldest grandchild, Jane Doust, the daughter of son Mervyn Ingram. Jane had been sent to Benenden School in 1949 at the age of twelve, from her home in Somerset. 'I was struck, even then, by the chaos in his writing room at the top of the house,' she told me. 'But Grandpa didn't go to school or university, so order was never instilled in him. On the other hand, his garden, which he had carved out of the farmland, was immaculate and as intriguing as Japanese gardens always are.'

Meanwhile Daphne had accompanied Alastair on assignments to British Army bases in Germany, Libya and Egypt after their marriage. During those years she gave birth to two children, Heather and Peter, in 1948 and 1950 respectively. After Alastair left the army in 1953, the family moved to Frame Farm in Benenden near The Grange. The farm was part of the land Ingram had bought in 1919. There, Alastair grew wheat and fruit and reared dairy cows and sheep. Among Heather's favourite memories of her childhood was wandering through

The Grange's cherry garden to bring milk and cream to her grandparents, hand in hand with her younger brother.

Ingram was especially fond of Heather, because she was an animal-lover and was interested in plants. 'I think I was the only one of twelve grandchildren who was allowed to enter Grandpa's greenhouse,' she recalled. 'It was overflowing with potted plants, and Grandpa knelt down and taught me the names of flowers.' Ingram also grew grapes, which he fed to Heather. More than four decades later she still recalls their exceptionally sweet taste, when thinking about her life at The Grange.

Summers at The Grange were an English idyll. On the front lawn the children played croquet until Grandma 'Flo' called them inside for home-made scones and cream. Whenever the grass in the cherry orchards grew too long, Alastair brought his sheep from the farm to graze them. Florence's hens pecked and strutted, as Ingram's dogs – Drongo, a Labrador, and Martin, a Norwich terrier, both named after birds – frolicked under the cherry trees.

Most summers Alastair, Daphne and the children decamped on an overnight sleeper train from London to the Bettyhill Hotel in the Scottish highlands with Collingwood, Florence and the dogs. While Heather and Peter walked Drongo and Martin or went swimming in the sea, Florence fished for salmon and Collingwood hiked around the nearby moors, looking at birds and wild flowers.

Daphne had a strong relationship with her father-in-law. She called him 'Cherry' and frequently drove him the 20 miles from Benenden to the seaside town of Hastings to buy fresh lobsters and fish. But one subject was out of bounds: Japan. By silent agreement, Ingram and Daphne refused to discuss anything to do with the nation that had taken away her freedom for more than three years.

Her experiences were raw and painful. Her repugnance towards the Japanese was not unusual among ex-prisoners of war and manifested itself in many small ways. She refused to buy Japanese cars or electrical products, despite their popularity in Britain after the 1960s. And she would have nothing to do with anything related to Japan, according to her children.

Daphne loved plants, and Ingram gave her many flowers from The Grange, including wild cyclamen, myrtle and cistus. But she never asked for a Japanese cherry tree. No matter how pure and beautiful the blossoms that flowered at The Grange, the trees' associations with that country held a darkness for her. English cherries, which Alastair grew in his orchard, were fine; they produced sweet, dark-purplish fruit and beautiful white blossoms.

While she learned to live with her past and built a contented new life for herself and her family, Daphne never talked about her time in Hong Kong. Whenever the subject was raised, she simply said, 'Nobody wants to know', even though her wartime memories shadowed her for the rest of her life. She occasionally took the train to Edinburgh to see Molly Gordon, her fellow nurse and internment-camp friend, who had been raped during the 'Black Christmas' atrocities. 'Molly was a lot older than me, rather like an older maiden aunt,' Daphne said. 'We never spoke of what had happened. We just shared a cup of tea and chatted.'

It was only when Daphne was ninety-three years old that she unburdened herself about the war for the first time, revealing little-known details about the Japanese takeover of Hong Kong in December 1941 and her three years of imprisonment in

Stanley Internment Camp. The prompt came from a journalist, Nicola Tyrer, who interviewed her in 2007 for a book about military nurses.

After Tyrer's book, *Sisters In Arms*, was published in 2008, she sent a signed copy to Daphne's care home in Winchester. Tyrer's book contained the stories of some of the 12,000 women who had served in the Royal Army Nursing Corps, which had been established in 1902 by Queen Alexandra. Unable to read the book because of deteriorating eyesight, Daphne sat quietly as her son, Peter, recited it aloud, reliving her past for the last time. Only then did Peter and his sister, Heather, understand why the bountiful cherry blossoms in Ingram's garden evoked such conflicting emotions.

'My mother seemed relieved that she was finally able to divulge her POW experience to society,' Peter told me. 'She seemed to have opened up to a journalist in a way that she never opened up to me or my sister.' Less than a year later, on 24 January 2009, Daphne died peacefully, aged ninety-four.

44. Cherries of a 'Traitor'

Daphne Ingram was a victim of the cherry ideology that had caused the deaths of millions. She was fortunate to have survived. Among the Allied powers, Britain had more POWs than any other nation, about a quarter of whom died while prisoners of the Japanese army.

The tentacles of this emperor-worshipping, self-sacrificing philosophy infected all aspects of Japanese life from the 1930s until 1945. This extended to the cherries, which the military government wanted to plant throughout Asia as a way to solidify the links between Japan and the lands it occupied overseas. In Kyoto the sixteenth successive Tōemon Sano, the son of the 'cherry guardian' who had worked with Collingwood Ingram to return *Taihaku* to Japan, told me the astonishing story of a wartime plan to plant as many as one million trees throughout China. The story had been largely unknown until the fifteenth Tōemon Sano published his memoirs in Japanese in 1970, and his son filled in the details for me.

The story began in the 1930s, when Japan was expanding its empire in China and when Count Kōzui Ōtani, the influential former leader of one of Japan's most powerful Buddhist sects, approached Sano with an audacious proposal. Count Ōtani, a cherry-tree connoisseur who had spent several years in London, asked Sano to grow 100,000 cherries in Kyoto, as the first part of an ambitious initiative to plant one million trees alongside railway lines in central Asia. These cherries, the count told Sano, would be planted at various places on the Silk Road that he had visited on three archaeological expeditions. The project tied in with the Japanese military government's expansionist plans to build a Pan-Asian railway. Several rail routes were planned, including one that would connect China and Turkey via Afghanistan, Iran and Iraq.

'My father was really serious about this project,' the sixteenth Sano told me. 'He started growing different varieties of cherries because the railway was going to be built through different climates. He was preparing Kurile cherries, originally from the islands north-east of Hokkaido, for cold areas, and varieties such as *Kanzan* and *Fugenzō* for milder areas.'

The 'million-cherries' project appeared to have support from the very top of Japanese society. Count Ōtani's sister-in-law was married to Emperor Taishō and was the mother of Emperor Hirohito. The count himself had been the head priest of Nishi Honganji, a subsect of the Jōdo Shinshu (Pure Land) form of the religion.

Count Ōtani and the fifteenth Tōemon Sano were advocates of Pan-Asianism – the idea that the Asian peoples should cooperate to counter Western imperialism. That philosophy was eventually hijacked by Japan's right-wing politicians and evolved into the concept of a Greater East Asia Co-Prosperity Sphere. During the war, Count Ōtani became an adviser to the government.

In 1940, impressed by the grandiose 'million-cherries' plan, Sano started buying large amounts of land in Kyoto's Funai District, about 40 miles north-west of the city. The land was large enough to plant the first batch of 100,000 cherry trees, or 10 per cent of the planned total. On the first 74 acres he started growing saplings.

To kick off the project, Sano, then aged forty-one, took 2,000 young *Yama-zakura* trees to Shanghai in March 1941 by ship. Nishi Honganji monks from Count Ōtani's sect were stationed there as missionaries. Together with senior Japanese military officers, the monks helped choose locations for the cherries, in places where battles had taken place during the Sino-Japanese War of 1937. Count Ōtani's intentions, Sano said, were 'to console the spirits of the Japanese soldiers who had lost their lives in battle' and to 'realise eternal friendship between the Japanese and Chinese'.

The plan was to plant several thousand cherries each year in Nanjing, Suzhou, Hangzhou and other Chinese cities. But by 1944, as the war turned against Japan, 'Count Ōtani's great ambitions and my efforts evaporated like a dream,' Sano wrote. As food became scarce within Japan, the 'million-cherries' initiative was abandoned and all but forgotten.

'One day in 1944, the police called me in and told me to convert my land for food,' Sano recalled. 'Flowers had become luxuries. Anyone who was devoted to cherry blossoms would be labelled a traitor. And such rumours had reached me.' He agreed to convert 10 per cent of his land to growing food. And he was forced to relinquish all the 100,000 trees, which were at various stages of growth. He donated 30,000 trees to anyone who would take them, including 300 flourishing trees to the nearby Utano Sanatorium in Kyoto, convincing the government that the hospital's tuberculosis patients would enjoy viewing the blossoms. Then Sano cut down the remaining 70,000 or so saplings himself, for use as fuel.

'I spent several days chopping, digging and pulling the trees out of the ground,' he remembered. 'It was heart-breaking, but I felt it was my responsibility, as I couldn't not sacrifice for the country any more.'

Still remaining in his Kyoto garden were seventy trees, each one a different variety. They were the rarest and most precious trees in his collection. Disregarding the authorities, Sano secretly kept them in a remote corner of the land, away from prying eyes, where his son proudly showed me some of the varieties that still grow there today.

'I believed in Japan's victory,' Sano wrote. 'But even if we lost, Japan would never vanish. I wanted to save these seventy varieties for the country, even though it meant risking my life.' Among these varieties was *Taihaku* – the tree that had gone extinct in Japan, and which Sano had painstakingly grafted twelve years earlier from scions sent back to Japan by Collingwood Ingram.

As for the trees that had been planted in Shanghai, the sixteenth Sano later heard that the Chinese authorities had uprooted them, as unwelcome symbols of Japanese militarism.

45. Britain's Cherry Boom

In a Britain racked by post-war austerity, the sight of cherry blossoms in full bloom was an uplifting experience. Few people associated these Japanese flowering cherries with Japan's conduct during the war, apart from former prisoners of war like Daphne Ingram. When Collingwood Ingram's book, *Ornamental Cherries*, was published in 1948, some varieties of cherry tree were already popular among the nation's plant lovers, largely because of Ingram's popularising efforts since the 1920s. Yet within a decade, cherries became a much more prominent part of the British landscape, planted and prized in gardens and parks, on roadsides and riverbanks. By the early 1950s a cherry-tree boom was under way. It quickly became self-sustaining.

'If readers of Captain Ingram's book are not inspired to write off immediately to their nurserymen with an autumn order, then they must indeed be dumb cold fish,' Vita Sackville-West wrote in a review of *Ornamental Cherries* for *The Observer* newspaper.

'No longer will they be satisfied with the gaudy *Kanzan*, the crude double-pink blossom with coppery foliage, ubiquitous in the gardens of bungalows, villas and suburbia. The enterprising planter can do better than that, and Captain Ingram will tell him how.' Sackville-West, Ingram's neighbour and the owner of Sissinghurst Castle, also pitched the book to local government councils, noting how the mass planting of cherries in America had become a tourist phenomenon.

The book was also well received outside Britain. This 'excellently conceived and well-executed monograph ... will doubtless remain the standard authority on this subject for many years to come,' the *American Nurseryman* said in its own review. The review was prescient, for as Sackville-West had predicted, local governments, plant wholesalers and gardens across the country began to import, plant and hybridise cherry trees. It was as if the trees themselves had been liberated by the peace and were now free to spread from garden to grove.

Amid that tide of enthusiasm, Ingram's personal imprint became particularly visible at what became Rosemoor Garden in the south-western English county of Devon. This followed a serendipitous encounter between the Ingrams and Rosemoor's owner, Lady Anne Palmer, who later became Lady Anne Berry. In the winter of 1959 the 40-year-old Lady Anne had been convalescing from a bout of measles at a hotel in the Spanish town of Algeciras, when she bumped into Florence and Collingwood. The Ingrams usually spent three months in this port every year during their old age, escaping the British winter.

In Lady Anne, Ingram found a kindred soul half his age. A wealthy descendant of Britain's first prime minister, Sir Robert

Walpole, she was a maverick horse- and bird-lover who had been educated at home, much like Ingram himself. Her half-sister, Lady Dorothy Mills, was an adventurous explorer who made a name for herself in the 1920s with her solo expeditions to Timbuktu, Haiti and Liberia.

'"Cherry" influenced my life more than any other individual,' Lady Anne told me. 'He literally opened my eyes to the wonders of nature, and to plants in particular.' The pair first bonded in Algeciras when the 79-year-old Ingram showed Lady Anne a vulture's nest. On other expeditions around southern Spain he pointed out plants and trees common to the area, 'awakening my latent interest in nature'.

Inspired by Ingram, Lady Anne decided that when she returned to England she would plant a garden on land in Devon that she had inherited from her father, the Earl of Orford. She named it Rosemoor Garden, with the intention of creating a 'mini-Wisley', in private homage to the RHS's showpiece garden south-west of London in Surrey.

'I took a Land Rover and a trailer and drove to Benenden,' Lady Anne recalled. 'We filled the trailer with a mountain of rhododendrons, primulas, polyanthus and other plants, including seedlings of *Kursar* and *Taihaku* cherry trees which "Cherry" had created. I was always careful to fill up any holes I may have made when I was digging under his supervision. He was very strict about this.'

Lady Anne was talking to me from Gisborne in New Zealand, where she was living with her second husband, Bob Berry, a fellow plant- and tree-lover who founded Hackfalls Arboretum in Gisborne. Her first husband, Colonel Eric Palmer, had died in 1980. At ninety-six years old, her hearing was deteriorating now, so she asked to continue our discussion via email:

> Cherry's knowledge of the natural world was amazingly rich, and he taught me a lot about the various plants that are particular to the coasts of Spain. He also talked a lot about cherry trees. If I hadn't met him, I would never have created Rosemoor Garden. He even drove back with me to Devon so that he could plant the 'loot' from Benenden, many of which are now large plants. When he gave me the *Kursar* cherry, I clearly recall him saying: 'By this plant, you must remember me.'

En route from Benenden to Devon, the pair stopped at Hillier & Sons nursery in Winchester, so that Ingram could introduce Lady Anne to Sir Harold Hillier, the nursery's owner and a fellow member of the elite Garden Society. Hillier & Sons supplied plants to Queen Elizabeth II and had created hybrids of some of the cherries that grew at The Grange.

Smitten by Ingram's passion, Lady Anne developed an interest in dendrology, the study of woody plants (trees and shrubs are 'woody plants') and their taxonomy. Over the following two decades she travelled to Japan, the United States and South America to add new varieties to her collection of cherries and other plants, with frequent stops at The Grange to see her mentor.

After Lady Anne moved to New Zealand in 1988, she donated the 40-acre Rosemoor site and buildings to the RHS. Sixty years later *Taihaku* and *Kursar* trees planted by Ingram are as robust as ever, Jonathan Webster, Rosemoor Garden's curator, told me. The garden now boasts more than forty types of flowering cherries, from the double-flowered, heart-shaped *Accolade* and the small, deep-pink blossoms of *Asano* to the greenish-yellow petals of *Ukon*. Rosemoor Garden also contains many rare rhododendrons from The Grange, together with their offspring.

Elsewhere in Britain, tens of thousands of cherries were planted between the 1950s and the mid-1970s, bringing colour, variety and a touch of Asian exoticism to the urban environment. The trees' popularity became evident in the names of streets, parks, pubs and restaurants. My husband's parents lived in a semi-detached house on Cherry Tree Avenue, for instance, in the picturesque Cheshire village of Lymm. The avenue had been named in the 1960s when the homes were built. At the top of their garden was a solitary cherry tree, probably planted by the builder. The avenue led off Cherry Lane, named a couple of generations earlier, when sweet English cherries grew there.

Check the map of any British town or village and there's almost always a 'Cherry' or 'Cherry Tree' avenue, close, park, road, street or way, mostly named during or after the 1950s, each containing a few hastily planted trees to justify its name. Predictably one road near the car-manufacturing plant in Sunderland operated by Japan's Nissan Motor Company was dubbed 'Cherry Blossom Way'.

From a virtual standing start in the 1920s, Japanese flowering cherry trees became a part of British people's daily lives within half a century. Their blossoms became an unmissable part of British springtime, showering parks and pavements with their tiny petals. The British always sought diversity, much as Ingram did. On tree-lined streets, local authorities selected different varieties, such as *Kanzan, Fugenzō* and *Umineko*, depending on the region. Other gardeners started to breed unique British varieties, such as *Accolade*, which was a cross between the Sargent wild-cherry species and the cultivated *Kohigan* variety. In Newcastle-under-Lyme in the west Midlands, for example, one residential street was lined with about fifty *Accolade* trees planted in the latter half of the 1960s.

As at Lady Anne's Rosemoor Gardens, hundreds of public and private gardens also planted an abundance of different cherries during the post-war boom. On the Batsford estate in the Cotswolds, for instance, the Japanese-themed gardens were originally developed by Algernon Freeman-Mitford, the British diplomat and author of *Tales of Old Japan*, and then sold in 1919 to Gilbert Wills, the 1st Lord Dulverton. The estate fell into disrepair during the Second World War. But when the 2nd Lord Dulverton inherited the estate in 1956, he started to replant the gardens with collections of ash, birch, magnolia, maple, oak and cherries. At Batsford, wild cherries such as Sargent and Fuji coexisted along with many of the flowering cherries that Ingram had introduced to Britain, including *Temari* and *Ichiyō* varieties that had come from the banks of the Arakawa River.

The cherry boom also permeated ordinary households. As well as varieties that Ingram had created, such as *Okame* and *Kursar*, smaller varieties became popular in urban gardens, including *Kiku-shidare*, a pink weeping cherry, and the late-blooming, wide-spreading *Shōgetsu*. Meanwhile in the countryside, where space was less of an issue, wide-spreading trees with pure-white blossoms, such as *Shirotae*, were planted along the driveways of country houses.

46. Ingram's 'Royal' Cherries

Britain's royal family also became infatuated with cherry blossom during the 1950s, in part because of the royal gardeners' links with Ingram. Four miles south of Windsor Castle, the Queen's principal weekend retreat, the 5,000-acre Windsor Great Park contains a variety of gardens, grasslands and woodlands, all managed by the Crown Estate. In the east of the park sits the Savill Garden, established in 1951 by Sir Eric Savill, the deputy park ranger at the time, with the support of King George VI.

In July 1948 Sir Eric wrote to Ingram, his horticultural friend, requesting 'some unusual varieties of cherry trees' for Windsor. Ingram immediately sent a batch of scions by parcel post. 'I have been guided in my selection by beauty and constitution rather than rarity. Most of the rarer varieties have rather poor constitutions,' he wrote.

Sir Eric asked Ingram for more trees in April 1949, saying that the white cherry blossoms that Ingram had displayed at a flower show were 'very much admired'. Ingram replied that Sir Eric was referring to the snow-white *Umineko*, the naturally hybridised cross of Ōshima and Fuji wild-cherry species that he had found growing at The Grange, and he duly dispatched some rooted cuttings.

When I visited the Savill Garden during a recent spring, the white blossoms of *Umineko* were in full bloom throughout the park. It was a stupendous sight, and in my mind's eye I saw the blossoms as Ingram did, grouped closely together like a flock of gulls sitting on top of a middle-sized tree. According to Mark

Flanagan, the Keeper of the Gardens until his death in October 2015, these trees were descendants of Ingram's original trees from The Grange.

Windsor Great Park gardeners have also planted numerous other varieties in the Savill Garden, including the autumnal *Jūgatsu-zakura* and *The Bride*, a large-petalled white blossom from Belgium. Strolling through the Valley Gardens, the 250-acre woodland part of Windsor Great Park, amid an abundance of azaleas, ancient oaks, rhododendrons and other trees and plants, a visitor can also see Fuji cherries, *Shirotae* and many types of *Matsumae* varieties from Hokkaido, Japan's northern island, donated by a cherry creator called Masatoshi Asari.

From the 1970s onwards, Flanagan and his predecessor, John Bond, planted cherry trees in the Valley Gardens. One particular tree stands out. Descending through the winding paths amid the bushes towards the lake below, the scenery opens out and a solitary cherry tree appears amidst other species. It's a *Taihaku*, and a descendant of one from The Grange. On the clear spring day of my visit, the blossoming canopy of large, pure-white flowers stood majestically beneath the blue skies, a perfect fit for the landscape.

In the north-east corner of the park sits the Royal Lodge, the residence for fifty years of Queen Elizabeth The Queen Mother, until her death in 2002. Japanese cherry trees were planted in the gardens of the Lodge, Flanagan told me. Although no records remain, these were probably the offspring of plants that Ingram gave to Sir Eric.

Wonderfully, it appears that the late Queen Mother was particularly fond of cherry blossoms. Staveley Road in Chiswick, a suburb in south-west London, is one of the capital's best-known cherry locations, with cherry trees lining both sides of the road. They are the *Kanzan* variety, planted in the 1920s. When the Queen Mother returned from Windsor Castle to Buckingham Palace in London after Easter, she would always ask her chauffeur to drive down Staveley Road. And her daughter, Queen Elizabeth II, continues this spring-time tradition.

47. The *Somei-yoshino* Renaissance

Britain's cherry boom coincided with a cherry renaissance in Japan, beginning in the 1950s. But, as before the war, British interest in variety contrasted with the Japanese passion for perfect uniformity. Once again Japan began to fixate on *Somei-yoshino*, the tree that had been at the heart of the cherry ideology of the 1930s and 1940s.

Most urban cherries had been wiped out in Japan during the war, leaving the cities bereft of colour and character. The revival began on a small scale as early as 1948, just three years after Japan's surrender, when 1,250 trees were planted in the war-burnt fields of Tokyo's spacious Ueno Park, one of the capital's first public parks. This had been a popular cherry-viewing, or *hanami*, site since the early seventeenth century, so it was natural that the government wanted to re-establish this tradition.

Within a decade, a nationwide rush to plant *Somei-yoshino* trees was under way. 'It was as if the clocks had been turned back fifty years,' wrote Akihito Hiratsuka, an author who has written extensively about the trees.

Following the Meiji Restoration in the 1860s, the central and local governments had planted *Somei-yoshino* as a symbol of Japan's emergence as a rising power. A hundred years later, in the 1960s, the conservative Liberal Democratic Party that has ruled Japan almost continuously since 1955 sought to make the cherry tree a recognisable global icon of the nation's rebirth, in much the same way that it promoted the beauty of Mount Fuji.

Hiratsuka called the decade between 1955 and 1965 a '*Somei-yoshino* bubble' because so many trees were planted by local governments, both to beautify land that had been bombed by the Americans and to revive the *hanami* tradition. From parks and riversides to tourist destinations, hundreds of thousands of *Somei-yoshino* trees sprouted from Kyushu to Hokkaido.

By 1964, when Tokyo hosted the Olympic Games, *Somei-yoshino* was again Japan's quintessential cherry. As the nation started to attract more foreign visitors, images of this one variety were used to promote the country's charms. Any lingering memories of the militaristic culture that the cherry had once represented were conveniently buried and forgotten. But the orgy of planting left much to be desired. In their rush to make the environment more beautiful, local governments often crammed hundreds of *Somei-yoshino* into tiny areas, with few gaps in between. Others were planted in open areas and left to grow freely, with minimal upkeep.

By the beginning of the twenty-first century about four out of every five cherry trees in urban areas in western Japan were *Somei-yoshino*. In other cities they accounted for nine out of every ten trees. Over the entire country, including non-urban areas but excluding the mountains, about 70 per cent of all planted cherries were the *Somei-yoshino* variety.

Meanwhile most of the flowering varieties that the 'cherry guardians' had saved were virtually invisible, planted as they were in the enclosed grounds of research organisations or in botanical gardens. Few of them grew where the general public could appreciate them easily.

As interest in the annual flowering of the blossoms re-emerged, the Japan Meteorological Agency began to publish official forecasts of the so-called 'cherry blossom front', or *sakura zensen*. In newspapers, on the radio and on television, the predictions of when the *Somei-yoshino* trees would begin to blossom in different parts of the country became a national event, eagerly monitored by *hanami* party-planners. When the reports were first produced in 1951, the agency studied and weighed the buds of the trees in specific areas throughout Japan, excluding tropical Okinawa and the frigid north of Hokkaido, where *Somei-yoshino* trees did not grow because of the climate.

The Meteorological Agency stopped these predictions in 2010, when private institutes developed more elaborate computer-generated forecasts. But it still announces the official 'first blossom of the year', based on studying five or six buds on each of three 'standard' *Somei-yoshino* trees in the grounds of the Yasukuni Shrine in Tokyo, where Japan's war-dead are venerated.

How can we explain the revival of *Somei-yoshino* so soon after the war, as well as the continued exclusion of other cherry varieties?

One explanation is that *Somei-yoshino* trees were simply convenient because they were fast-growing, cheap to buy, easy to maintain and beautiful. And the Japanese buyers and sellers of the cherries — mainly government officials and nurserymen

– had all come of age in the twentieth century when *Somei-yoshino* was the archetypal tree and diversity was ignored. The dominance of this one variety was normal and unquestioned by them.

In the same way that only one cherry variety was planted, Japan approached reconstruction from its wartime ruins with a narrow-mindedness that echoed its breakneck drive to modernise and then militarise from the late nineteenth century onwards. Although the post-war reconstruction era began with a determination to learn lessons from (and never to repeat) the past, the nation soon started moving again along a solitary track.

The government's goal this time was to become an economic powerhouse, and by following an aggressive industrial policy, Japan raced from destitution in 1945 to virtual domination two decades later, as the world's second-largest economy after the United States. It was a momentous achievement. It was no surprise that the *Somei-yoshino* revival accompanied Japan's singular and unswerving pursuit of economic glory. The cherry clone had kept pace with Japan's modernisation and militarisation eras before and during the war. And during the post-war rebuilding period, the *Somei-yoshino* cherry had been given a second wind.

For some people, the *Somei-yoshino* blossoms were favoured simply because they flowered at the same time each April, when schools and universities held graduation and entrance ceremonies. 'They developed an image as a sentimental flower that marked crucial points in the lives of Japanese people,' Tōru Koyama, a researcher at the Flower Association of Japan, told me. Because of that, he said, the flower's future popularity was assured.

But not everyone agreed. One vocal critic of *Somei-yoshino*'s dominance was the sixteenth Tōemon Sano. The *Somei-yoshino* variety has only existed for 150 years at most, Sano stressed to me. Given the 2,000-year-plus history of Japan's cherry trees, the monotone scenery of the twenty-first century is an historical exception rather than the norm. And he continued, 'Despite being a small country, Japan's climatic conditions differ dramatically from one place to another. Each part of Japan produces unique foods, reflecting nature's diversity.

'Cherry blossoms are innately diverse by nature. Each region has its own cherries. Some bloom early, others bloom later. In the past, whenever they blossomed, people would plant rice. And if the blossom was late, the people took it as a sign of possible late frosts, so they would wait to plant the seeds.

'The spreading of this single variety, *Somei-yoshino*, disregards such differences, and it is wrong to call this variety the "standard" cherry tree. It's like forcing people to speak a standard form of Japanese, ignoring regional dialects.'

Sano sighed deeply and finished his thoughts.

'Wherever you go these days, the blossoms are all the same. To me, the singularity of *Somei-yoshino* makes Japan a uniformly boring country.'

PART SEVEN

CHERRIES OF RECONCILIATION

48. A Garden of Memories

The publication of *Ornamental Cherries* in 1948 was the high point of Collingwood Ingram's cherry research. He was sixty-seven years old. Afterwards, Ingram wrote a handful of articles about Japanese cherries for national newspapers and magazines. He continued entering plants from his collection in horticultural society competitions – and won numerous awards well into old age. He remained a fount of wisdom about cherries, happy to discuss them and to give away grafts and seedlings of favourite varieties to anyone who asked.

But the passion with which he had pursued his study of the trees before the war dimmed in the 1950s and 1960s. Perhaps this was a natural tailing off or he didn't want to offend Daphne. Or perhaps it was because by the end of the 1950s the five Japanese cherry enthusiasts and friends with whom Ingram had corresponded had sadly died: Seisaku Funatsu in 1929, Professor Manabu Miyoshi in 1939, Masuhiko Kayama in 1944, Aisaku Hayashi in 1951 and Count Nobusuke Takatsu-kasa in 1959.

Ingram was also distressed that many of the cherries he had planted at The Grange in the 1920s did not live as long as he had expected. He told several friends, 'The trouble with Japanese flowering cherries in Britain is that most only live for forty or fifty years.' While wild cherries have a longer lifespan, varieties created by humans are less resilient.

One wonders, too, whether Ingram's disillusionment with Japan affected his love for the nation's iconic flowers. He never expressed his thoughts about the emperor-worshipping cherry ideology that had led Japan into the Pacific War, but he was certainly critical of Japan's rapid industrialisation both before and after the war and wrote that the country had 'changed beyond recognition'. 'That a people who were once incomparably the most artistic race in the world should have prostituted their heritage and permitted the destruction of some of their most beautiful scenery for industrial development is past belief,' he stated.

During the 1950s and 1960s, with the publication of *Ornamental Cherries* a success, Ingram systematically waded through volumes of material that he had collected or written decades earlier. His intent was to round out his life with a couple of autobiographical books that would interweave his love for travel with specific examples of the birds and plants that he had found in exotic locales.

Birds once again played a larger role in his life. At weekends, Ingram would often take his eldest granddaughter, Jane Doust, to nearby Romney Marsh to observe birds and listen to their different calls. He was particularly concerned by the decline in the number of birds in Britain since the early twentieth century, including one of his favourites – the nightingale. At one time the nightingale had nested in his garden, but by the late 1950s it was rare to hear one sing within earshot of The Grange.

Ingram attended the annual International Ornithological Congress three times during the 1950s. At the conference held in Finland in 1958, he met the son of Count Nagamichi Kuroda, the founder of the Japan Ornithological Society, whom he had met in 1926 in Japan. The following year Kuroda gave Ingram the title of honorary member of the society. He kept the letter from Kuroda for the rest of his life.

In 1966, aged eighty-five, Ingram published *In Search of Birds*, based partly on articles he had written since he was a young man. Writing for 'the average bird lover', he criticised the jargon- and statistics-filled prose of many fellow naturalists, and instead wrote simply about the birds he had seen, from Alaska to the West Indies.

Four years later, in 1970, he published *A Garden of Memories*, a book about plants that he had found while travelling or that he had hybridised at Benenden. Among these were two flowers that he named after the village: *Rubus* 'Benenden' and 'Benenden Blue' rosemary. A chapter in the book about cherries was intended as a supplement to *Ornamental Cherries*, written twenty-two years on.

As in most of his books, the frontispiece opposite the title pages of *In Search of Birds* and *A Garden of Memories* was a drawing by Ingram of a bird sitting on a cherry-tree branch inside a rectangle. 'He often told me that this logo represented his lifelong love of both ornithology and botany,' Moira Miller, Ingram's housekeeper for five years, told me. Miller had lived at The Grange with her first husband, Robin Tomsett, and their son, Fraser. In his books Ingram also included a simple but elegant drawing of a cherry with its stalk. The 'C' stood for 'Cherry' and 'Collingwood', while the stalk stood for 'Ingram'.

Even in his eighties and nineties, Ingram was extraordinarily agile and energetic, according to Sibylle Kreutsberger, the joint

Birds and botany: Ingram's frontispiece

head gardener with Pamela Schwerdt at Sissinghurst Castle Garden from 1959 to 1990. Ingram visited Vita Sackville-West at Sissinghurst periodically in the years leading up to her death in 1962. He continued visiting the gardens, which contained a *Taihaku* tree that he had given Vita, for the next eighteen years. 'He always brought puppies with him and would have happily climbed a fence in his nineties if he had to,' Kreutsberger told me. 'He once said to me, "Florence is ninety-four. She can only play golf half a day nowadays."'

Besides cherries, Ingram had always enjoyed propagating rhododendrons, more than thirty of which he considered 'really worthwhile'. In his later years he found them easier to take care of than cherries and built a formidable and prize-winning collection.

He had three other valuable collections at The Grange that had little to do with trees or plants. Like Bertie, his eldest brother, Ingram owned hundreds of Japanese *netsuke* (miniature ivory or wood carvings), *inro* (lacquered seal-cases) and *tsuba* (sword guards). Most were kept in airtight walnut display cabinets in his cold and cluttered attic. He had collected them when he was in his twenties, buying most of them in London before their value appreciated. Also on show were stuffed birds, pressed flowers, antique books and sheaves of animal prints.

It was Ingram's intention to donate these collections to the British Museum after his death. So in 1973 he invited Lawrence Smith, a specialist in Japanese artefacts, to The Grange to look at them. Smith reported that he always took at least two thick sweaters when he visited, so that he could withstand the temperature in the attic. Smith, who later became senior keeper of the Department of Japanese Antiquities at the museum, thought Collingwood and Florence were a charming old-fashioned aristocratic English couple, albeit somewhat eccentric. Whenever he went to the house, he was always served lunch. 'He [Ingram] sat at one end of a long table and she sat at the other end,' Smith told me. 'I had to sit in the middle, and they were shouting at each other because they were a bit deaf. She said: "Do you think Mr Smith wants another helping of pudding?" and he shouted back, "I don't know, why don't you ask him?" They were very formal. Amid the uproar, their big liver-and-white setter dog would try to steal food from my plate.'

When Ingram picked up Smith in his car at the train station in Staplehurst, seven miles north of Benenden, the first thing he said was: 'Would you keep a lookout in case you see something coming?' 'He clearly couldn't see, but he didn't care what people thought,' Smith recalled. Ingram drove everywhere in second gear and guided himself by the white lines in the centre of the road.

Indeed, his dangerous driving became a Benenden legend. But no one dared tell him to stop, because he had been a magistrate and was the most important person in the village. Finally, one day, a concerned police officer lay in wait at the main gates of The Grange and pulled Ingram over.

'Excuse me, Captain Ingram, can you read that car's number plate?' the officer asked, pointing at a nearby vehicle.

Ingram peered and said, 'What car?'

The policeman was apologetic, but took away his driving licence on the spot. He was ninety-seven.

'He was incredibly angry,' Moira Miller recalled, laughing. 'He shouted, "I've been deprived of my licence. I want it back."' After that, Moira's husband, Robin, took over the driving.

Despite his failing eyesight, Ingram was determined to publish further ornithological musings that he had omitted from *In Search of Birds*; he was keen to expound what he called his 'controversial' beliefs, many of which did not 'coincide with those of other ornithologists'. Unable to find a publisher, he paid for the book's publication and sold it privately in 1978. Called *Random Thoughts on Bird Life*, it was his last work, bringing to an end a body of writing about cherries, plants and birds that stretched back more than seventy years.

That same year Ingram welcomed a special guest from America, with whom he had been regularly corresponding about cherry blossoms. Roland Jefferson, chief botanist at the

US National Arboretum in Washington D.C. and the most famous African-American botanist in the US, was renowned for his research on the Potomac trees and the capital's annual cherry-blossom festival. While most trees on the Potomac were the *Somei-yoshino* variety, some varieties at the National Arboretum came from Ingram's collection at The Grange, including *Hokusai* and *Taihaku*. By the 1970s, Ingram told Jefferson, his extensive cherry collection at The Grange was 'sadly depleted, either from neglect or by removal to make way for other trees and shrubs'.

Jefferson didn't care. 'Mr Ingram was very famous in the United States as a cherry tree expert, so I had always wanted to meet him,' he told me from his home in Honolulu. 'After we talked for about an hour, sitting on a bench in The Grange, he guided me around his garden and pointed out the remaining varieties.'

Collingwood Ingram and Roland Jefferson at The Grange, 1978

49. A Peaceful Death

On the evening of 29 November 1979, Florence Ingram died at the age of ninety-seven. The couple had been married for seventy-three years. Even though Ingram paid little attention to his family when they were young, he and Florence had grown close during their quiet years in Benenden, and their time in southern Spain together every winter. 'They were very different characters, but they were always great friends,' reported Ruth Tolhurst, who was born in a cottage in The Grange's grounds and had known them all her life. 'He was mad on hunting, whereas she was more interested in fishing. And if she was late coming home from fishing, he got a bit cross if Florence wasn't at the dinner table at 7 p.m.'

Florence's death left Ingram grief-stricken, his routine shattered. A few days later, Tolhurst went to see him and found him sitting mournfully at the long table in the dining room, eating breakfast. 'He said: "What am I going to do in this big house on my own?" So I said: "Why don't you get another puppy to train?"' Soon afterwards, Ingram bought Noddy, an energetic Norwich terrier. For the rest of Ingram's life, Noddy — named after a tropical seabird — never left his master's side.

Ingram developed a regular routine. Early in the day he would stroll around the garden with a cane, affectionately examining and feeding the trees and plants, using compost from great heaps in the garden. After a bowl of porridge with fresh cream at 9 a.m., he would return to the garden. At 1 p.m. promptly it was lunch, always typical British fare: fresh fish, lobster or roast beef with vegetables from the garden, a glass of wine and a sweet pudding. After an afternoon nap, he would shut himself in the attic to sort out his papers, not leaving the room until dinner.

On 30 October 1980 Ingram celebrated his 100th birthday with a small party at The Grange for his family and close friends. The owner of the fish shop in Hastings brought lobsters for his most regular customer, which Moira Miller cooked for all the guests. Among the multitude of birthday cards was one from Queen Elizabeth II, which is sent to all centenarians.

In an article to mark his centenary, Roy Lancaster, one of Britain's most celebrated plantsmen, noted that since Ingram's birth a 'cavalcade of famous plant hunters had bowed out from the Asian stage', while Ingram was still tending his own garden decades later. 'Like all true plantsmen, he regards his plants as his children, his collection as his family,' Lancaster said after a visit to The Grange. 'Each is a character, each has a history – a cork oak grown from an acorn collected whilst on a boar hunt in Portugal, a lodge pole pine collected as a seedling from a bog in Alaska, a climbing hydrangea from the rain forests of Chile. And if it had not been for his untiring effort in introducing new or long-lost cherries, our gardens would be a lot poorer.'

To further recognise Ingram's achievements, Michael Zander, a botanist from Kew Gardens, made a list of the trees and plants at The Grange, mapping the places in which each grew. Sadly, only twenty-three varieties of cherry tree were identifiable. On the map, Zander recorded several wild-cherry species. They included *Yama-zakura*, Fuji cherry and Sargent cherry, one of which was the original plant Ingram had collected in Japan in 1926. The map also showed that many of his favourite cultivated cherries were still present, including *Kursar, Hokusai, Imose, Shirotae, Taoyame* and *Taihaku*. Zander did not mention the total number of cherry trees at The Grange. There were several trees of each variety, and others whose names were unknown.

In all, it appears there were between forty and fifty trees, certainly far fewer than the 120 varieties that had at one time grown in the garden, but enough to retain the impression of a well-loved cherry orchard. Zander's inspection of The Grange was part of a larger project to catalogue woody plants in cultivation in Britain, many of which were under serious environmental stress by 1980. 'Plants, which once seemed plentiful, have become a precious commodity and a valuable resource,' Zander wrote. 'The days of the great plant hunters and extensive nurseries may be gone, but thankfully many of the gardens they inspired and the plants they collected are still with us.' Fortunately Ingram's cherry trees were in no danger of extinction, even if many of those at The Grange had died. All the varieties he had collected were growing somewhere in Britain or abroad.

After his centenary, Ingram arranged to send seedlings of rhododendron and primrose that he had raised in his greenhouse to his close friend Alan Hardy, owner of the Sandling Estate north of Benenden. Ingram sensed that his life was drawing to a close.

The spring of 1981 was to be his last. The bright-pink *Kursar*, a variety that Ingram had created forty years earlier, bloomed early in the season. Soon afterwards *Yama-zakura*, which had so impressed him in Koganei fifty-five years earlier, opened. *Hokusai* – the tree at The Grange that had first ignited Ingram's interest in cherries in 1919 – followed. Then came *Taihaku*, the flower that he had returned to its homeland. As

Ingram paced slowly around his garden with Noddy at his heels, it was a trip down memory lane, filled with happiness and more than a little sorrow, as the petals fell over the course of two months.

From time to time Lady Anne Berry, his Rosemoor Garden friend, visited to talk about plants, and the past. 'The last time, he had just turned one hundred years old,' she recalled. 'I drove him to lunch at a local pub, but he was failing fast by then.'

In early May, Ingram complained of feeling unwell and was confined to bed. He didn't visit the garden again. As he weakened, nurses stayed overnight in his bedroom to monitor his condition. His daughter, Certhia, was able to tell him that the swallows, which nested each year in the porch of The Grange, had arrived. For the first time in years, Ingram was unable to attend the RHS's annual Chelsea Flower Show but Alan Hardy took along a rhododendron that Ingram had created. It was the first time a centenarian had entered this competition. Over the years, Ingram won more than 100 awards from the society, including the prestigious RHS Veitch Memorial Medal in 1948 and the Victoria Medal of Honour in 1952.

'Days before his death, Alan was digging up a lot of things in the garden, but Cherry sat up in his bed one day and said: "When I get over this illness, I want them back,"' Charlotte Molesworth, a longtime Benenden resident, told me.

On the evening of 19 May 1981, Collingwood Ingram passed away. It was a peaceful and painless departure. In the garden of The Grange, the late-blooming *Imose* was shedding its pale-pink petals. It was the cherry that Ingram had found in the Hirano Shrine in Kyoto in 1926 and had introduced to Britain for the first time.

The life of 'Cherry' Ingram – the English 'cherry guardian' – ended as the blossoms that he had loved throughout his adult life waved silently in the wind outside his room. It was the day of the full moon.

50. The Grange after Ingram

Collingwood Ingram's funeral service at St George's Church was attended by an overflowing crowd of relatives, friends, Benenden residents and horticultural colleagues from around the UK. Afterwards, a small group of family and close friends gathered at The Grange. Later, Ingram's ashes were interred in St George's churchyard, next to Florence. His obituary featured in all the major British newspapers. *The Times* called him an 'indefatigable traveller in search of plants' and said that he had 'what is probably the finest collection of (ornamental) cherries in the world.'

'"Cherry" was the last of a breed of wealthy Edwardian gentlemen of leisure who began as amateurs in a specific scientific discipline and then became experts,' Peter Kellett observed, summing up Ingram's life. 'He loved hybridising everything: cherries, rhododendron, primulas, even crab-apples. He once told me he was a hybrid himself – between a bird and a cherry.'

The day after Ingram's death Lawrence Smith from the British Museum travelled to The Grange to collect the *netsuke*, *inro* and *tsuba*. In accordance with Ingram's will, some 1,100 items were donated to the then-Department of Oriental Antiquities, now the Department of Japanese Antiquities.

Smith was just in time. Noddy, Ingram's dog, had chewed several books and materials that Ingram had left behind. 'A rare book of cherry tree paintings, called *Sakura Zuhu*, written by Professor Manabu Miyoshi, would have been damaged if I had come a few days later,' Smith recalled. Ingram had bought the book, which depicted 112 different cherry trees, during his last trip to Japan. It was among his favourite possessions.

After Ingram's death, The Grange was sold to Martin and Judith Miller, the entrepreneurial founders of popular antiques guides. They hired Charlotte and Donald Molesworth in 1983 to restore and maintain the fast-deteriorating garden. The Molesworths were landscape garden designers who had bought the cottage next to The Grange, in which Sidney Lock, Ingram's gardener, had lived.

A couple of years later The Grange was sold on to the rock musician and producer Alan Parsons and his then-wife, Smokey. Parsons, a recording engineer on The Beatles' *Abbey Road* album and founder of the Alan Parsons Project, immediately converted the old scullery into a state-of-the-art recording studio. It became a rural haven for London-based musicians and producers, including Paul McCartney.

The Parsonses gave the Molesworths a free hand to continue the garden renovations. They planted a glade of Japanese maples, along with irises, peonies, lilies and cherries.

Yet on the night of 15 October 1987, six years after Ingram's death, a lot of that hard work was undone. Hurricane-force winds and rain, the worst in decades, battered southern England and France, causing more than a dozen deaths, including one in Benenden. The Grange was badly hit, with more than 100 trees torn down by the storm, including Atlantic cedars, pines, walnuts and an 80-foot tree from Chile.

A huge fallen oak tree also blocked the driveway, and seventeen flowering cherries died. They included a rare crimson-blossomed *Carmine* cherry that Ingram had grown from seedlings given to him in the early 1930s by Frank Kingdon-Ward, a British plant-hunter. Despite the destruction, the Parsonses and the Molesworths decided to open The Grange to visitors the following April for the National Garden Scheme's annual charity day, an event that Ingram had supported since the late 1920s.

Soon after the storm, the Parsonses sold The Grange. The house and garden passed through three further owners over the next thirty years. Today it is an assisted-living home for adults with learning disabilities.

51. Home and Abroad

Ingram's garden, the cradle of British cherry blossoms, no longer takes the breath away. But the collection that he began to accumulate in 1919 lives on throughout the United Kingdom. The cherries that Ingram gathered and grafted are now an integral part of the British countryside and the urban environment.

Japanese flowering cherries can be found throughout Britain's parks and gardens – it is now hard to find places where they do not flourish. Britain cannot get enough of cultivated cherry blossoms, according to Nick Dunn, the owner of Frank P. Matthews Ltd, one of the country's largest suppliers. In recent years, demand at garden centres has continued to rise and the company sells about seventy different cherry varieties.

Yet the market for cherries has changed somewhat since the post-war boom. Then, many local councils planted cherry trees along roadside pavements. When the trees' shallow roots grew, they sometimes dislodged the paving stones. As government budgets tightened, maintaining cherry trees became a lower priority. Now it is in public and private parks and gardens, on university campuses and in urban and rural gardens that cherry trees abound.

They have become increasingly popular features at many British visitor attractions. The RHS gardens at Wisley, Surrey are a case in point. Cherries have grown there for more than 100 years, but it wasn't until the 1980s that they were widely planted. Between 1983 and 2017 Wisley gardeners planted 157 cherry varieties, from the pink *Accolade* to the purple *Yae-murasaki*. One popular addition in 2011 was the *Collingwood Ingram* variety. The tree was raised in 1979 from a set of *Kursar* seedlings by Ingram's friend, Robert de Belder, owner of the Arboretum Kalmthout in Belgium. And the RHS has planted a colonnade of 100 *Somei-yoshino* trees in front of a new entrance hall at Wisley, to create a sense of seasonal drama for its million-plus annual visitors.

Kew Gardens, Britain's most famous garden, has also featured Japanese flowering cherries since the 1920s and has added many new varieties since the 1990s. Tony Kirkham, head of the Arboretum at Kew, oversaw the construction of a 'Cherry Walk' containing numerous cultivated varieties at the south-west London site. Adjoining it is a row of thirty doubled-flowered *Asano* trees, all direct offspring from the tree that Ingram found at the foot of Mount Fuji.

Ingram would have been pleasantly surprised to see the *Asano* trees at Kew Gardens. He would be even more astounded by the 350 *Taihaku* trees that grow in the gardens of Alnwick Castle in Northumberland, near England's northern border with Scotland. It is the world's largest collection of 'Great White' cherries. They are all descendants of trees that once grew at The Grange.

First planted in 2008, the Alnwick *Taihaku* are increasingly popular as they reach full growth, scattering clouds of white blossoms onto a carpet of bright-pink tulips and 50,000 yellow and white daffodils. The gardens adjoin Alnwick Castle, which was used as the location of Hogwarts School of Witchcraft and Wizardry in the *Harry Potter* film series and for some scenes in the *Downton Abbey* TV drama.

According to chief gardener Trevor Jones, the *Taihaku* project evolved from a plan to create an imaginative modern garden for the castle, after Ralph Percy, the 12th Duke of Northumberland, inherited the dukedom in 1995. His wife, Jane Percy, Duchess of Northumberland, has loved cherry trees since she was a child and wanted to create something that had never been done before.

'When the pure-white blossoms come off the trees, I want it to look like a snowstorm,' the duchess told me as we ate sandwiches and scones in a treehouse restaurant in the gardens. 'I want to take people's breath away. If you have varieties, you prolong the blossom season, but you don't get that incredible Wow-factor.'

As dusk descended on the gardens one springtime evening, the duchess led a procession of Northumberland pipers and local families down through the sloping *Taihaku* grove to a small pond. Each family had sponsored one or more trees in memory of someone who had died. As they walked, small groups would break off from the procession to pray at or hug their tree, or to sit and sway in silence on wooden swings. As the Alnwick Chamber Choir sang, the bereaved families lit candles in paper lanterns, which drifted slowly across the darkening water.

'These cherries are a sign of hope,' observed Canon Paul Scott, vicar of St Michael's Church in Alnwick, at the *Taihaku* ceremony. 'They're one of the first flowers to blossom after a long winter, so I see them as symbols of new life and new possibilities.'

The duchess knew little about Collingwood Ingram when she decided to plant only one variety, *Taihaku*, and she was excited to hear the story of its return to Japan. She has no regrets with her choice: '*Shirotae* [a popular white-petalled variety] are lovely, but *Taihaku* are magnificent'.

Elsewhere in Britain, towns and cities vie to show off the diversity of their cherry trees. Harrogate in Yorkshire boasts its famous cherry-tree walk on The Stray. In north-west England the gardeners at Tatton Park, Ness Botanic Gardens on the Wirral, Arley Hall and Dunham Massey all tout their cherries. So, too, does the Dartington Hall estate in south Devon, which has featured *Somei-yoshino* and *Taihaku* trees since Beatrix Farrand, an American landscape designer, planted them in the 1930s.

Meanwhile, Batsford Arboretum and Keele University, near Newcastle-under-Lyme, hold National Plant Collections of

Japanese flowering cherries, recognised by the British charity Plant Heritage for their efforts to preserve cherries. And in another ambitious initiative to spread cherries throughout Britain, a private project endorsed by the Japanese and UK governments involves the planting of more than 3,000 trees throughout the UK as 'a visible symbol' of the two nations' relationship. To begin with, the trees will be planted in Royal Parks in London.

In most countries with temperate climates, the cherry's ubiquity means it is hard to escape its blossoms in spring. Scores of cities boast a favourite cherry-blossom park, street or festival. And there is usually a specific reason why they were planted there in the first place. Take Germany, for example. The post-war capital of Bonn beautified its Old Town – *Altstadt* – by planting sixty *Kanzan* cherries on both sides of a historical street called the Heerstrasse. When the blossoms bloom each April, their distinct pink flowers create a canopy over the street, which is now known as Cherry Blossom Avenue. In Hanover, in the 1980s, the city planted 110 cherry trees, which it received from its twin city, Hiroshima. Meanwhile, Berlin has planted more than 9,000 flowering cherries along the former 'death strip' through which Eastern Bloc residents once tried to escape to the West. To pay for the trees, a Japanese television station, TV Asahi, collected more than $1 million from its viewers in 1990, following the collapse of the Berlin Wall.

The Italian capital of Rome, too, has had a cherry-blossom tradition since the Japanese government donated about 2,500 trees in 1959 to create a cherry path, or 'path of Japan', around an artificial park called Lake Park in the south of the city. It is now the country's most renowned cherry-viewing location, although there are also thousands of trees in the streets and parks of Milan, the nation's northern industrial centre.

And so it continues. The US West Coast city of Seattle holds an annual festival in appreciation of the 1,000 cherry trees given by the Japanese government in 1976 to commemorate 200 years of American independence. Philadelphia, St Louis and the New York borough of Brooklyn, among other US cities, also hold cherry-blossom celebrations. From the Sydney cherry-blossom festival in Auburn Botanic Gardens and the Jinhae Gunhangje festival in South Korea, to the Meghalaya festival in north-eastern India and the Vancouver Cherry Blossom Festival, cherry blossoms are celebrated throughout much of the world every spring. Most festivals highlight the flowering cherries that have been popular over the decades.

And in Britain over the last twenty years, a new cherry type has been gaining popularity because of its originality and beauty. It is called *Matsumae*, a generic term for 116 double-flowered varieties created in Japan since 1959 by Masatoshi Asari.

These *Matsumae* varieties were first propagated in England by Britain's foremost cherry-blossom expert, Chris Sanders, in the 1990s. Sanders and Asari didn't meet until 2017, almost a quarter-century after Sanders had first set eyes on Asari's creations. But each would help to inject new life into the cherry-blossom world in both Japan and Britain and, in so doing, would burnish the legacy of Collingwood Ingram.

52. The Next Generation of *Sakuramori*

As a thirteen-year-old boy growing up near the southern Hokkaido port town of Hakodate in late 1944, Masatoshi Asari listened quietly to his elder brother, Shōichi, talking about the British and American prisoners of war who lived in desolate wooden camps a few miles from their home. In the middle of winter, said Shōichi, who worked for the military as a volunteer, he often witnessed the stick-thin, frost-bitten foreigners sitting next to a road-construction site, chewing small pieces of boiled squid for their lunch to stave off hunger.

Among the POWs' tasks were the removal of seaweed and barnacles from the hulls of visiting military ships and the loading and unloading of coal and cement. As they walked to and from work, the men repeatedly sang, in Japanese, the patriotic Aikoku March. Prison guards beat the men with rifle butts if they couldn't remember the complicated words.

Many of the prisoners didn't survive: 174 of them died in the Hakodate camps between 1942 and 1945. Royal Air Force Leading Aircraftman Lawrence Richardson, from north Wales, died in January 1943 of acute inflammation of the colon, leaving a grieving widow, Ethel. Cyril Breach, a Royal Air Force Volunteer Reserve airman from south-west London, passed away the following month of acute inflammation of the kidneys. Royal Artillery gunner John Derbyshire, an unmarried twenty-four-year-old from Lancashire, died of beriberi in May 1943. His parents, Thomas and Alice, left an inscription on his gravestone in the Yokohama War Cemetery

Yokohama Nursery Catalogue from 1926–27,
Ingram ordered extensively from the company during this period

Umineko, named by Ingram after the black-tailed gull

Kursar, cultivated by Ingram from wild cherry species,
in Chris Lane's nursery, Kent

The loathed *Kanzan*, planted across Britain in the 1960s
and at Ingram's daughter's school, much to his horror

The Grange, spring 2015. Now an assisted-living home,
forty *Matsumae* cherry trees were planted to mark the millennium

Ingram aged 99 at The Grange in 1980, underneath the cherry blossoms

Taihaku at The Alnwick Garden, Northumberland,
where a memorial ceremony is held each spring

'Reconciliation cherries': *Matsumae* varieties
in a private nursery at Windsor Great Park

where most of the dead prisoners of war from Hakodate were eventually buried. It read: 'Treasured still with love sincere. Just a memory but oh, so dear.'

The conditions were far worse than those at the Stanley Internment Camp in Hong Kong, where Collingwood Ingram's eventual daughter-in-law, Daphne, was a POW. Other camps in Japan had bad reputations but the Hakodate camp was considered a 'notorious hell-hole even amongst the Japanese'.

After the war, few Japanese people wanted to recall or be reminded of their wartime experiences. Masatoshi Asari, who as a young man became an elementary schoolteacher and amateur historian, was an exception. With a small group of volunteers, he researched the Hakodate POW camps, their conditions and the prisoners who had lived and died there.

He learned about the British, American, Australian and Dutch prisoners brought to the camps from the Malay Peninsula, Java and Singapore. He went to the British Embassy in Tokyo, where the consul-general helped him find the testaments of former POWs. He interviewed nurses in Hakodate who had visited the camps with the *kenpeitai* (secret police), and he talked to children who had climbed through holes in the fence to give the POWs cooked potatoes and corn balls.

Conditions in the camp were dreadful. During the long, freezing winter nights in their wooden huts, the scabies-ridden men huddled on straw mats under thin, lice-infested blankets near open-trench toilets. Their three meals every day consisted of watery gruel, with the occasional fish head or vegetable. When a stray cat or dog entered the camp, it became part of the stew. The guards tortured prisoners for minor infractions of the camp rules.

The prisoners' work was equally difficult. Besides labouring in dry docks and at construction sites, the men worked in mines and factories that supported the Japanese war effort. When Hokkaido's main newspaper mentioned the prisoners, the articles talked about their 'lack of pride and shame' and their 'selfishness'. 'They are more concerned about their family's safety than their country,' one article said. 'It is something we can't understand from our Japanese *bushidō* spirit.' The newspaper reflected the widely held view that it was a disgrace for a soldier to be captured alive by the enemy. The Japanese government had signed the Geneva Convention on the treatment of prisoners of war, but did not ratify it, because the military strongly opposed it.

Asari's findings, which he published in three booklets, sickened him. He could never get the images of the Allied prisoners of war out of his head, and resolved to try and atone somehow for his country's barbarism.

Later in life Asari became one of Japan's leading cherry-blossom experts and the creator of the spectacular group of cherry varieties called *Matsumae*. He was one of a select group of post-war Japanese plantsmen who knew about and admired Collingwood Ingram. He was in his twenties when he first read about Ingram in *Sakura*, a book published in 1937 by Ingram's friend, Professor Manabu Miyoshi. Miyoshi noted that Ingram had pointed out that Japan's flowering cherries were not in good health.

Asari had photocopied and devoured every page of Ingram's classic book, *Ornamental Cherries*, in the late 1950s, after borrowing it from a fellow botanist. 'All the botanists and researchers in Japan owned *Ornamental Cherries*,' he told me. 'We were very surprised that such a person existed in England.

He was creating new cultivars while we Japanese weren't. He pointed out things that we didn't even notice.'

Thousands of miles from snowy Hokkaido, Britain's top twenty-first-century *sakura* expert, Chris Sanders, trod a completely different path from Asari, his Japanese *sakuramori*, or cherry guardian, counterpart.

The son of a market gardener, Sanders left school at sixteen in 1960, completed the RHS Master of Horticulture degree at Pershore College of Horticulture in Worcestershire, and in 1966 joined John Hill & Sons. At that time the 400-acre nursery in north Staffordshire offered about seventy varieties of cherry tree for sale, but as Britain's cherry boom started winding down in the late 1960s, a dozen or so well-known varieties came to generate the bulk of their sales.

It became difficult to justify selling rarer cherry varieties for which there was little demand, so Sanders, as a young manager of the nursery, simply stopped stocking them. It was a similar scenario at nurseries across the country. But as Sanders learned to appreciate the subtle distinctions of different cherries, he started to regret his hasty actions. Without nursery sales, there was a danger that some lesser-known varieties would eventually disappear in Britain.

Sanders decided to follow Collingwood Ingram's path – to protect and save the cherries. On his days off and on holidays he became a plant-hunter: driving slowly around Staffordshire villages and towns, he would peer into people's front gardens, looking for rare varieties that were no longer on sale in England.

One day he spotted a light-pink *Edo-zakura* 10 miles from his home and convinced the owner to give him scions. Elsewhere he found *Daikoku, Taki-nioi, Taoyame* and other cultivars. 'One rare variety was growing on a street in a housing estate near Newcastle-under-Lyme,' Sanders recalled. 'I stopped and took photos of all these trees, and knocked on people's doors. It's a good job I wasn't arrested, because people must have wondered what I was doing.'

Over the years, in the garden of his Eccleshall home and at Bridgemere Nurseries, where he worked after leaving John Hill & Sons in 1982, Sanders collected and grafted saplings of more than 100 cherry varieties. Most were offspring of flowers that Collingwood Ingram had grown at The Grange.

Joining Sanders on many of his plant-hunting expeditions was another nurseryman called Chris – Chris Lane. The two Chrises had met in 1980 at a plant-identification course at Harlow Carr gardens in Harrogate, Yorkshire. As their friendship grew, Lane and Sanders started searching for rare flowering cherries at nurseries and gardens throughout Europe.

Later, when Sanders retired from Bridgemere Nurseries in 2002, he asked Lane to keep his collection of about 100 cherry varieties safe at Lane's Witch Hazel Nursery in Sittingbourne, Kent, just 30 miles north of Benenden. Today Lane's cherry collection is the world's largest, with close to 350 varieties. The only comparable collection is at the Flower Association of Japan's Yūki experimental farm in Ibaraki Prefecture, north of Tokyo.

In 1992, John Bond, one of Britain's top plantsmen, had received an invitation from the Flower Association of Japan to lecture about British cherry blossoms, especially about the Japanese flowering varieties that had been spread far and wide because of Collingwood Ingram's influence. Bond was the influential Keeper of the Gardens at Windsor Great Park for twenty-seven years. He advised the Queen in the 1980s on the restoration of Frogmore Garden, the royal family's hideaway in Windsor Great Park, and consulted with the Queen Mother about her garden at the park's Royal Lodge. When Bond returned from Tokyo, he made it a goal to import new varieties to add to the cherry collection at Windsor Great Park, and he left the selection to Chris Sanders, his longtime friend.

After referring to the English-language edition of the *Manual of Japanese Flowering Cherries* and considering the local climates, Sanders selected twenty-five strains of a newly bred cherry, Masatoshi Asari's *Matsumae*, for Bond to introduce to Windsor Great Park. At the time *Matsumae* varieties were unknown in Britain and were still rare in Japan, apart from in the northern island of Hokkaido. 'Almost all the old cherries had been introduced into the UK by Ingram. So I wanted to introduce something new,' Sanders said. At Sanders's suggestion, Bond wrote to Asari in January 1993, asking to buy his cherries.

Bond's visit to Japan and his ambitious cherry plan came at a complicated time in Britain and Japan's political relationship. The killing and suffering that the cherry ideology unleashed during the war had provoked predictable resentment in Britain against Japan. It usually simmered beneath the surface, until a specific event prompted an outcry. In October 1971, during the visit to the UK of Emperor Hirohito and Empress Nagako, a Japanese cedar tree planted by the emperor at Kew Gardens was pulled out of the ground and a former prisoner of war prominently placed a wreath on the monument to Britain's war dead in Whitehall that read: 'With vivid memories of the treachery and inhumanity of the Japanese'.

Similarly, after the emperor died in January 1989, the British media was filled with articles examining his culpability for the war and concerning the treatment of prisoners. In response, a group led by Keiko Holmes, a Japanese widow living in London, invited twenty-six former British prisoners of war and two widows to Japan in 1992 as a way to atone for the past. Holmes had become involved with POW activities after seeing a memorial stone for British prisoners who had worked down a copper mine in Mie Prefecture, east of Osaka, where she had grown up.

The Japanese government was apathetic about these private visits at first, saying that they had resolved the POW issues after the war by paying reparations to around 200,000 ex-POWs from fourteen countries based on the San Francisco Peace Treaty of 1952. But as criticism against Japan mounted in Britain, a group of ex-prisoners sued the Japanese government in 1993 for personal compensation. The government could no longer remain indifferent and started funding Holmes's 'reconciliation trips.'

53. Cherries of Reconciliation

In January 1993, an embossed envelope from Windsor Great Park dropped through the letter box at Masatoshi Asari's timber-built home in southern Hokkaido. Opening it carefully, Asari saw that the letter was from John Bond, a man he had never heard of, asking to buy some of his *Matsumae* cherries. To receive a letter on royal stationery was beyond Asari's imagination. To think that he – the son of a farmer from a remote frontier island – had created something fit for the Queen of England was like a dream.

In his poignant reply to Bond, the 62-year-old Asari wrote that he wanted to donate to the park the cherries that he had created, without charge. It was his way, he said, of expressing his 'sincere condolences and regrets to those who lost their lives during the war, and their bereaved families'. And he continued:

> I have devoted myself to the breeding of blossoming cherry trees, or *sakura*, for the past forty years and have been rewarded by seeing many different varieties created. It was an honour to receive your request for various kinds.
>
> I have a strong personal wish to present *sakura* trees to the people of the UK. Some fifty-one years ago, the Japanese armed forces invaded your territories and killed and injured many soldiers and civilians. I have never forgotten this historical fact.
>
> I sincerely hope that the *sakura* trees being sent from Japan will be carefully tended and raised, and that some day the blossoms will give pleasure and consolation to all those who see them, including those bereaved families of the war dead.

'After all my interviews and research about the POWs, I wanted to make amends,' Asari told me. 'We never repaid the British who came to Hakodate during the Meiji era and taught us about their water supply, drainage systems and shipbuilding technology. Instead, we treated them cruelly during the war.'

On 5 February 1993 Asari sent an enormous parcel to Bond containing the scions of fifty-eight different *Matsumae* cherry varieties that he had created. Thankfully, a European Union ban on cherry imports did not come into effect until later in 1993, so the package was not opened by agricultural inspectors. Sanders grafted the scions and planted them at Bridgemere Nurseries, with two skilled nurserymen, Neil Bebbington and Stephen Brookes, and the fifty-six varieties that grew were then planted at Windsor Great Park.

These cherries now grow in a private plant nursery in a corner of the park. On a visit I made there one spring, varieties such as *Matsumae-hanasomei*, *Matsumae-fuki* and *Matsumae-sasameyuki* were in full bloom. Their petals of white, pale-pink and red painted a breathtaking picture. Some of the *Matsumae* trees, as well as about thirty other flowering cherries, also adorn the Savill Garden and the Valley Garden within Windsor Great Park. And the royal family enjoy them in Frogmore Gardens, their private garden, too. A complete set of *Matsumae* cherries also grows at Keele University, and another in Chris Lane's nursery in Kent.

And, thanks to a serendipitous stroke of fortune, *Matsumae* trees also blossom each spring in the garden at The Grange in Benenden.

Quentin Stark, a young gardener at Windsor Great Park, was present when Asari's *Matsumae* saplings started arriving there from Chris Sanders's Bridgemere Nursery in the late 1990s. Stark planted one entire set of the young trees, which were then about five feet tall, in the private nursery. He and others planted another set of the cherries throughout Windsor Great Park. But they still had forty *Matsumae* trees left. Stark didn't know what to do with them.

One day, the answer hit him. Stark was familiar with Collingwood Ingram's legacy. He had grown up in Penshurst, 20 miles west of Benenden, and he had read *Ornamental Cherries*, which he had bought at a second-hand book sale at the Savill Garden. 'As the *Matsumae* cherries were new varieties, I thought that The Grange, being the residence of the world authority on cherries, would be an appropriate home for them,' he told me. His boss agreed.

It was eighteen years since Ingram's death, but on a visit to Benenden in 1999, Stark drove to The Grange and knocked purposefully on the front door. The owner at the time, Linda Fennell, answered. Stark had a proposal: the Crown Estate wanted to give forty *Matsumae* varieties to The Grange.

Fennell immediately agreed. They decided to plant the trees to celebrate the millennium. And so on Thursday, 2 March 2000 residents of The Grange gathered outside to watch as the forty trees were planted around the garden. It was exactly eighty years since Ingram had first seen two cherry trees blossom in his new garden.

One cannot help but marvel at how these 'reconciliation cherries', developed in Hokkaido by a POW historian who was also a fan of Ingram, found their way to The Grange without their creator's knowledge.

Asari has never visited Britain. And he didn't know, until I told him, that his creations were growing in the garden where Ingram had spent so much time with Daphne, his POW

Masatoshi Asari at his home in Hokkaido, November 2018

daughter-in-law. Neither did Chris Sanders, or members of the Ingram family. All were astonished by the discovery.

'It's miraculous,' Asari told me. 'I feel that the spirits of Collingwood and Daphne Ingram must have invited the cherries to The Grange for forgiveness and closure, to give them both peace of mind. Mr Ingram truly loved *sakura*. He valued life, and he wanted to make cherries a shared world heritage.'

As we chatted, Asari grew philosophical about his cherry mania. The many varieties of cherry trees are the consequence of ceaseless transformation by nature, and by man, over more than 2,000 years, Asari noted. 'You need to look at their long history when you view today's *sakura*. We had a rich cherry diversity in the Edo period and then rushed towards singularity after the Meiji Restoration, both in *sakura* and in society. We focused solely on modernisation and militarisation, which subsequently led us into war. That was a mistake. Now, if we want to construct a healthy and peaceful future, we have to learn from history, confronting our past mistakes.'

Only then did I realise that Asari's preoccupation with cherries, and his desire to unlock the history of the POWs, sprang from the same roots. He believed that diversity enriched nature and humanity alike; and that its absence had led Japan to disaster.

Somehow, Asari thought, his unwavering desire to atone for Japan's mistreatment of POWs had guided *Matsumae* cherries to Britain and then to The Grange. And his most fervent hope, he told me, was that *Matsumae* varieties that now thrive in Ingram's former garden serve as symbols of rebirth, redemption and restitution, offering comfort to Daphne, the Ingram family and all former POWs.

On 14 May 2017, surrounded by the blossoming *Matsumae* cherries that he had created, Masatoshi Asari shook hands with Chris Sanders in Matsumae Park in Hokkaido. Twenty-four years after their relationship had begun, they were meeting for the first time. Asari wore a light-coloured Western suit, while Sanders wore a blue Japanese *happi* coat.

It was an emotional moment. Asari and Sanders had been in contact since Sanders grafted Asari's *Matsumae* cherries in 1993. In Hokkaido on that cool spring day the pair unveiled a granite 'monument of friendship between Japan and England for cherry-tree appreciation'.

'I had always dreamed of my cherries blossoming beautifully in England, so it was an honour to finally meet the person who grafted and spread them,' Asari told me.

As of 2016, nineteen *Matsumae* varieties out of the fifty-six cultivars growing in England had received the RHS's Award of Garden Merit – a major accomplishment for new varieties. Two new cultivars created by Asari also now grow in Hokkaido. One is a double-flowered variety with pinkish-white petals, called the *Bond* cherry. Asari raised it from a seed in his garden and named it to honour John Bond, whose letter had prompted his gift to Windsor Great Park. The second cultivar is called the *Chris* cherry. Grafted from a plant that grew from the seeds of an *Amanogawa* cherry variety that Asari found in Hakodate, it is fittingly named for Chris Sanders.

Collingwood Ingram's meeting in Tokyo in April 1926 with his cherry-guardian mentor, Seisaku Funatsu, paved the way for the return of *Taihaku* to Japan and for Ingram to spread Japanese flowering cherries throughout the British Isles. Now it was the turn of the next generation of Japanese–British *sakura-mori* to carry on that tradition.

EPILOGUE

—

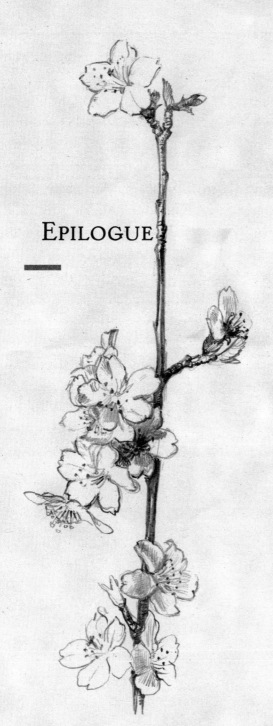

1·4·23

Prunus incisa

54. Millennia Trees

Deep in the mountains of central Japan, the ash-tinted petals of one of the oldest cherry trees in the world attracts 600,000 visitors a year. Neo village, where this majestic tree lives, has a population of 2,500, and when the tree blossoms, the rural roads are blocked by tourists for miles around.

The tree is an *Usuzumi-zakura*, which means 'pale-coloured ink tree', because its single-petal flowers turn shades of grey before falling. About 1,500 years old, it is one of three cherries that the Japanese government has designated as 'national monuments'. The others are an 1,800-year-old *Jindai-zakura* cherry in Yamanashi Prefecture, south-west of Tokyo, and a 1,000-year-old *Miharu-Takizakura* cherry in Fukushima Prefecture, due north of the Japanese capital.

None are in pristine condition. After all, they've been growing for a millennium or more. Yet the three trees are living memorials to Japan's long cherry tradition. While many cherry varieties disappeared in the cities of modern Japan, small groups of passionate cherry-lovers in the countryside have refused to let old and rare trees expire. Collingwood Ingram would have been delighted.

On a chilly day in April 2017 I took a taxi from Gifu, a city about 220 miles west of Tokyo, to Neo in the wooded foothills of Mount Nogo-Hakusan. Awaiting me there was Takeshi

Ōhira, a 43-year-old tree surgeon whose forestry company has a contract to keep the *Usuzumi-zakura* alive. With additional help from local volunteers, the tree is surviving against all the odds.

There is a big difference between saving a 50-year-old and a 1,500-year-old cherry tree. At first glance, the *Usuzumi-zakura* looked to be on its last legs. It is so decrepit that the inside of its trunk and branches are hollow, and parts of its bark have been stripped off. Unless it is protected, the tree will collapse and die. Indeed, it has already experienced numerous crises, surviving only because of human support. For instance, the tree's spreading branches have been propped up with more than thirty poles made from cedar trees.

In 1910 heavy snow cracked the trunk in many places, and by 1948 the tree's condition was dire. Only radical surgery by

The 1,500-year-old Usuzumi-zakura in Neo village

Toshiyuki Maeda, a Gifu dentist and amateur botanist, saved it. Maeda and a team of seventy-three workers replaced 238 of the tree's roots, which were infested with white ants, with younger ones from other trees, in effect sustaining the elderly tree with young root grafts. A decade later, the 56-foot-high tree was battered in a typhoon and only properly recovered in the early 1970s.

Ōhira, who sees himself as the tree's 'father', checks the *Usuzumi-zakura* carefully each week. In November the hard work begins. As the region braces itself for the harsh winter, Ōhira and his crew carry out *yuki-tsuri* (snow-lifting), a Japanese technique that aims to protect trees from heavy snows by supporting the branches from above. First, eleven cedar poles are set up in a circular pattern around the tree, and between fifteen and thirty straw ropes are attached to the top of these poles. Sitting in a conical bucket winched up by a large crane, a gardener ties the ropes to the cherry branches, working downwards. Over two long days at least 200 branches are suspended, taking the weight of the snow off the fragile tree. At the foot of the tree, Ōhira dumps eight tons of *susuki*, or Japanese pampas grass, to protect the roots and nourish the soil.

The *Usuzumi-zakura's* condition has deteriorated since it was designated a national monument in 1922. Nonetheless it is still an overpowering sight in full bloom, albeit smaller and thinner than it once was. The spread of its branches from one side to the other has shrunk from about 160 feet to 90 feet in the past century, as lengthy older limbs die and fall off. Equally disturbing is the fact that fewer blossoms appear each year. Yet provided there is no natural disaster – no typhoon or earthquake, no extreme snow or violent windstorm – Ōhira is confident that the tree will be fine for the next sixty years or so.

It's a big 'if'. To protect the *Usuzumi-zakura* for generations to come, the prefecture has cloned the tree, and a dedicated volunteer group is growing its children, collected from the tree's seeds within a fenced-off area. After hundreds of seeds have been found and cleaned, they are placed in a fridge to let them 'experience winter' – an essential natural germinating process for some species that have evolved in sharply seasonal climates – before being planted in pots. More than eighty saplings were growing in 2018. They are expected to flower in about a decade.

Looking at the *Usuzumi-zakura* and listening to Ōhira's enthusiastic explanations of the care that sustains it, and the plans for its future, gave me hope. Much like Ingram ninety years earlier, I had been worried that Japan's cherry-blossom diversity was in danger. In the cities, where *Somei-yoshino* trees dominate, that is true, with a few exceptions. In the countryside and in the mountains there are plenty of diverse cherries, and plans to grow more. The most ambitious cherry project is in the coastal city of Iwaki, one of several communities devastated by the earthquake, tsunami and nuclear disaster of 2011. So that was where I headed on my next visit to Japan.

55. The Great Wall of Cherry Blossoms

When the Great East Japan Earthquake struck at 2.46 p.m. on Friday, 11 March 2011 the lives of millions of people in north-eastern Honshu, 150 miles north of Tokyo, changed in an instant. Registering at 9.0 on the Richter scale, the earthquake triggered tsunami waves that travelled as far as six miles inland. More than 19,000 people died, and one million buildings were partly or completely destroyed. The torrent of water also smashed into the Fukushima Daiichi nuclear plant on the coast, causing the biggest nuclear incident since the Chernobyl disaster in 1986.

Concerned about radiation, the government evacuated more than 100,000 people from their homes in the areas around the plant. Many still live in temporary accommodation outside the no-entry zone that surrounds the stricken nuclear reactors. The region's economy was devastated. Lingering fears about this 'invisible enemy' also meant that fewer visitors arrived to enjoy the region's mineral-rich *onsen*, or hot-spring baths. Demand for its seafood, vegetables, fruit, rice and trees plummeted.

In the summer of 2016 my 17-year-old son, Kenji, travelled to Iwaki, a city about 35 miles south of the Fukushima nuclear plant, to work as a volunteer at a hotel that was offering rooms to others who were helping Iwaki residents to recover from the disaster. What he saw, and the tales he heard, opened his eyes to the scale of the catastrophe. He wanted to return as soon as possible to help and to see new friends, so we travelled there together late the following year.

Since 2011 cherry trees, a symbol of life and rebirth, have been planted in great numbers near Fukushima, in memory of those who died and to help resurrect neighbourhoods washed away during the tsunami. Of all these ventures, the Manbon Zakura initiative stands out, a multi-year project to plant 99,000 wild cherries in the hills west of Iwaki.

When I arrived in Iwaki, more than six years after the disaster, the residents were still sad and angry. While the earthquake and tsunami were unavoidable natural disasters, there was an undeniable sense that the nuclear accident could have been prevented.

Tadashige Shiga, a 68-year-old Iwaki businessman and the Manbon Zakura's project leader, was blunt. 'We were told the nuclear plant was safe. It wasn't, so we're also to blame for believing the government's lies,' he told me. 'We have done irreversible damage to our home town, as well as to people's minds and emotions. We're leaving an unmistakable negative legacy for generations. It's everyone's fault. We need to leave something that will give people hope.'

On undeveloped woodland owned by sixty Iwaki residents, volunteers have already planted thousands of cherry-tree saplings. They are all wild cherries, such as *Yama-zakura*, *Edo-higan*, *Ōshima-zakura* and Sargent cherries – species that Ingram loved. Most have been sponsored by individuals and companies. Besides Shiga, the biggest individual donation has come from a renowned Chinese artist, Cai Guo-Qiang, now based in New York.

Cai's connection to the area stemmed from the late 1980s, when he first moved from China to Tokyo at the age of thirty and struggled to survive. Unable to find a gallery that would display his unusual paintings, he moved to Iwaki, where he held his first exhibition. Cai's projects are spectacular because he paints with gunpowder, which he ignites to scorch the canvas. Among the friends he made while living in Iwaki was Shiga. So after the tsunami, Cai visited Japan to help Shiga with the Manbon Zakura project.

'The cherries we're planting now are cherries of repentance and hope,' Shiga said, stressing the last three words. 'It'll take about fifty years for them to look really gorgeous and to attract lots of people, and I won't be around to see them. But we want to leave a new legacy. We want our grandchildren to say, "Grandma, Grandad, look! Why are there so many cherries here?" Then the grandparents can tell them about the earthquake, the tsunami, the nuclear disaster, and how we tried to lift people's spirits by planting cherries.'

For Cai, the ambitious project was an opportunity to repay Iwaki for its help when he was a penniless artist. 'If we plant 99,000 trees, you'll be able to see them from space,' he said. 'It would be the Great Wall of Cherry Blossoms. Wouldn't that be wonderful!'

Indeed. I felt awed by the scale of the projects I had encountered, with the Neo venture protecting a primordial tree, and the Iwaki plantings projecting a brighter future. These grassroots schemes contrasted sharply with the monolithic conformity of most cherry projects in Japan, in which only *Somei-yoshino* trees were planted. The joy was that there were a further three community-based programmes: two in Tokyo, one further south. Some of my Japanese countrymen and women, I was delighted to learn, were fighting the *Somei-yoshino* tide.

For instance, in Tokyo, some three hours south of Iwaki, the residents of Adachi Ward are resurrecting the once-famous 'five-coloured cherry trees' of the Arakawa River, which Ingram admired with his fellow cherry expert, Seisaku Funatsu, in 1926. Tragically, all the trees there disappeared shortly after the war ended. What varieties survived did so only because of Funatsu's foresight in sending scions to a relative for safekeeping, decades earlier. Those varieties now thrive in a government cherry-tree conservation forest west of Tokyo. Other trees from the Arakawa were sent to Washington D.C. in 1912, as part of a plan to grow them on the banks of the Potomac.

Cherry trees in bloom in Washington D.C.

With some local government help, the Adachi residents had planted more than 450 young cherries of forty-nine varieties along about three miles of riverbank by 2016. There they grew in complement to 120 trees of eighteen varieties that were planted in the 1990s along a smaller section of the river. Those 120 trees, along with many more, came from Washington D.C., mostly from the National Arboretum. Most were descendants of saplings sent to the US capital in 1912. A few, including the *Taihaku* variety that Ingram repatriated to Japan and the *Hokusai* variety that he named, were offspring of Ingram's collection at The Grange.

I found a similar mood of confidence at Koganei, the district in Tokyo where cherries at the once-celebrated cherry-viewing spot had virtually disappeared during the 1930s. When Ingram visited in 1926, more than 1,400 *Yama-zakura* trees graced the canal, which delivered water to Tokyo residents. By 1965 the number of trees had halved, and their condition had deteriorated because of pollution and road-widening projects. The erection of a safety fence next to the canal added to Koganei's demise.

Yet since 2010 eager residents have been re-creating the famed cherry avenue, planting *Yama-zakura* saplings, building paths along the canal and removing overgrown trees. Within a decade both Arakawa and Koganei should once again be prime places within the capital to see a range of diverse cherry blossoms.

There was another heart-warming tale on the Izu Peninsula, south of Tokyo. There, in the tiny town of Kawazu, the discovery of an early-blooming cherry, now known as *Kawazu-zakura*, has invigorated the economy. A natural hybrid of the Ōshima cherry and the Taiwan, or *Kanhi-zakura* cherries, this long-lasting salmon-pink variety was found by Katsumi Iida in 1955 and has since been cloned countless times. More than a million people now visit Kawazu each February and March to walk under the 8,000 trees, which stretch along a river almost two miles inland from the Pacific Ocean.

Now that I was nearing the end of my four-year cherry-blossom journey of discovery, I found myself thinking about the theories of variety and diversity again. Whenever I thought about the cherries' extraordinary history, I was struck by the tragedy that had allowed Japan's naturally varied canvas to be painted over with the uniform hues of a single variety. As Japan learned through bitter experience, a society with tunnel-vision is an unstable one. 'If you stick with singularity, you may get results quickly,' Asari told me. 'But neither a flower nor a society can evolve with richness and vitality if everything and everyone is the same.'

Nor could I help recalling Ingram's 1926 appeal to Japan's elite to prevent the decline of cherry varieties. In my heart, I wanted to believe that grassroots efforts in Japan to protect and encourage cherry variety would ultimately become the mainstream. In my head, I wasn't too sure. Japanese society had been drifting to the right, mirroring global trends, and Ingram's warning remains a potent message for the Japanese. But it is also a global message. Rereading Ingram's writing, it was all too clear how much the preservation of cherry varieties and species – indeed, of all plant species – meant to him.

Ingram was never more passionate than when he wrote about nature's bounties; never angrier than when he wrote about mankind's destruction of nature. An environmentalist long before the term became popular, he deplored plant-hunters who selfishly uprooted all the flowers they were seeking. He also resisted what he called 'rarity snobbism' – the syndrome of being influenced by a plant's scarcity rather than by its beauty. And he railed against mankind's failure to appreciate the Earth's diversity and to protect its fragility.

Even in the 1920s Ingram was concerned about the planet's future and the consequences of unfettered economic growth. One journal entry that he wrote during a visit to Sumatra has even greater resonance now than when it was written more than ninety years ago:

> Progress, improvement, development – call it what you like – is rapidly reaching even the remotest corners of the globe. Wherever modern man comes into contact with Nature, he leaves a disfiguring mark. As his

numbers multiply, so the fundamental beauty of the universe decreases.

The passing of beauty and romance from this world is, to me, a source of endless regret. When the Victoria Falls have been harnessed and Spitzbergen turned into a teeming coalfield, it will be time to think of another planet.

A stone's throw from The Grange, in the graveyard of St George's Church, gusts of wind whipped fallen blossoms into the air, late one Sunday afternoon. After a visit to Sissinghurst Castle Garden, I had driven to Benenden to look for the graves of Collingwood and Florence Ingram.

They weren't easy to find. The ground was damp and many of the gravestones were covered by decaying leaves. I pushed them to one side and tried to read the inscriptions.

After half an hour or so I had scrubbed clear the corner of one stone and saw the outline of what I had been searching for. One word: 'Ingram'. A little more cleaning and the inscription was clear: 'William Alastair Ingram 1913–1975 and his loving wife Daphne Anne Ingram 1914–2008'. My heart beat a little faster. Close by was a smaller tablet: 'Florence Maude Ingram 1881–1979'. Adjacent was a weathered headstone with a barely visible engraving: 'Collingwood Ingram 1880–1981'.

I felt light-headed. Below me were the ashes of a man I had never met, but whom I had come to know; a man whose cherry-blossom footsteps I had followed from Britain to Japan and back. What now, I thought? My journey was complete, but I didn't want to leave quite yet. I imagined Ingram standing by my side, little Noddy at his feet, explaining to me the legacy of every cherry at The Grange.

I considered the fate of the cherries that Ingram had loved, created, protected and saved from extinction. Not just *Taihaku*, but wild-cherry species and old and new cultivated varieties alike. Many were cherries that once flowered only in Japan. Now they bloomed around the world, in arboretums and parks, along city streets and riverbanks and in millions of suburban gardens. I thought about his passion, his family and the generations that he had inspired. Ingram had helped to change the face of spring. He had spread beauty around the world and helped to create a shared treasure – the cherry blossom – for all to enjoy.

The daylight faded. The wind picked up. Rain threatened. I took a long, last look at Ingram's headstone, dusted off my hands and headed home to London.

Appendix A

Key cherry varieties

Taihaku / *Prunus* (Sato-zakura Group) 'Taihaku' / Great white cherry

Somei-yoshino / *Prunus x yedoensis* 'Somei-yoshino' / Tokyo cherry

Kanzan / *Prunus* (Sato-zakura Group) 'Kanzan', syn. 'Sekiyama'
   Note: although 'Sekiyama' is mostly used in Japan, 'Kanzan' is always used in Europe

Asano* / *Prunus* (Sato-zakura Group) 'Asano'

Kursar** / *Prunus* 'Kursar'

Hokusai* / *Prunus* (Sato-zakura Group) 'Hokusai'

Daikoku* / *Prunus* (Sato-zakura Group) 'Daikoku'

Shirotae / *Prunus* (Sato-zakura Group) 'Shirotae', syn. 'Mount Fuji'

Umineko** / *Prunus* 'Umineko'

Okame** / *Prunus* 'Okame'

Fugenzo / *Prunus* (Sato-zakura Group) 'Fugenzo', syn. 'Shirofugen', *Prunus serrulata* f. *alborosea*
   Note: the cultivar name 'Albo-rosea' is used in Japan, but not in Europe

Matsumae varieties created by Masatoshi Asari / *Prunus* 'Matsumae' / some UK trade names include Chocolate Ice and Fragrant Cloud

*Named by Collingwood Ingram

**Created by Collingwood Ingram

Key wild cherry names

Yama-zakura / *Prunus jamasakura,* syn. *Prunus serrulata* var.
   spontanea / Japanese mountain cherry

Mame-zakura / *Prunus incisa* / Fuji cherry

Ōyama-zakura / *Prunus sargentii* / Sargent cherry

Kanhi-zakura / *Prunus campanulata* / Taiwan cherry, Bell-flowered
   cherry, Formosa

Ōshima-zakura / *Prunus speciosa* / Ōshima cherry

Edo-higan (also *higan-zakura*) / *Prunus spachiana* f. *ascendens,* syn.
   Prunus pendula f. *ascendens* / Spring cherry

Appendix B

Cherry blossom viewing locations

Cherry blossoms can be viewed at thousands of locations around the world. This list includes some of the most popular places where their beauty can be enjoyed.

United Kingdom

Alexandra Gardens, Cardiff

Alnwick Castle, Northumberland

Arley Hall, Cheshire

Batsford Arboretum, Gloucestershire

Brogdale Collections, Kent

Dartington Hall, Devon

Dunham Massey, Greater Manchester

Greenwich Park, Kew Gardens, Regent's Park and St James's Park, London

Keele University, Staffordshire

Maxell Gardens, Telford, Shropshire

National Botanic Garden of Wales, Carmarthenshire

Ness Botanic Gardens, Cheshire

RHS Garden, Rosemoor, Devon

RHS Garden, Wisley, Surrey

Savill Garden and Windsor Great Park, Surrey

The Meadows, Edinburgh

The Stray, Harrogate, Yorkshire

Tatton Park, Cheshire

Westonbirt Arboretum, Gloucestershire

Europe

Alster Lake, Hamburg, Germany

Amstelveen, Amsterdam, the Netherlands

Arboretum Kalmthout, Belgium

Bispebjerg and Langelinie Park, Copenhagen, Denmark

Heerstrasse, Bonn, Germany

Herbert Park, St Stephen's Park and National Botanic Gardens,
   Dublin, Ireland

Japanese Garden, Hasselt, Belgium

Jerte Valley, Extremadura, Spain

Kungstradgarden, Stockholm, Sweden

Lake Park, Rome, Italy

Mauer Weg and Glienicke Bridge, Berlin, Germany

Petrin Park, Prague, Czech Republic

Powsin Botanical Garden, Warsaw, Poland

Roihuvuori, Helsinki, Finland

Romberg Park, Dortmund, Germany

Americas

Arnold Arboretum and Charles River, Boston

Branch Brook Park, Newark, New Jersey

Brooklyn Botanical Gardens, New York

Botanical Garden of Curitiba, Brazil

Central City Park, Macon, Georgia

Descanso Gardens, Lake Balboa and Lacy Park, Los Angeles

Fairmount Park Horticulture Center, Philadelphia

High Park, Toronto

Japantown, San Francisco

Missouri Botanical Garden, St Louis

Potomac River and US National Arboretum, Washington D.C.

Queen Elizabeth Park, Vancouver

State Capitol State Park, Salem, Oregon

University of Washington, Washington Park Arboretum and Seward Park, Seattle

Japan

Arashiyama, Daigoji Temple, Kyoto Botanical Gardens and Kyoto Imperial Palace, Kyoto

Expo 70 Park, Japan Mint and Osaka Castle Park, Osaka

Hirosaki Park, Hirosaki City, Aomori Prefecture

Imperial Palace moat, Shinjuku Gyoen National Garden, Sumida Park and Ueno Park, Tokyo

Kawazu River, Kawazu Town, Shizuoka Prefecture

Matsumae Park, Matsumae Town, Hokkaido

Mount Ontake Natural Park, Bungo-Ōno City, Oita Prefecture

Mount Yoshino, Yoshino, Nara Prefecture

Sakurayama Park, Fujioka City, Gunma Prefecture

Tama Forest Science Garden, Hachioji City, Tokyo

The Flower Association of Japan's Yūki farm, Yūki City, Ibaraki Prefecture

Three national treasures: *Usuzumi-zakura*, Motosu City, Gifu Prefecture; *Jindai-zakura*, Hokuto City, Yamanashi Prefecture; *Miharu Takizakura*, Miharu Town, Fukushima Prefecture

Asia-Pacific

Alishan National Scenic Area, Taiwan

Auburn Botanic Gardens, Sydney, Australia

Cowra Gardens, New South Wales, Australia

Gucun Forest Park and Shanghai Botanical Gardens, Shanghai, China

Gyeongpodae Pavilion, Gangneung, Gangwondo, South Korea

Jinhae Gunhangje festival, Jinhae, South Korea

Kings Park and Botanic Garden, Perth, Australia

Longwangtan Cherry Blossom Park, Dalian, China

Shillong, Meghalaya, India

Tortoise Head Gardens, Wuxi, China

Yangmingshan National Park, Taiwan

Yeouido Park, Seokchon Lake and Kyung Hee University, Seoul, South Korea

Yuyuantan Park, Beijing, China

Notes

PART ONE

1. *Family Ties*

p.15   **Within the hat**: 'My friends, in feather and fur' by Lady Ingram, *The Windsor Magazine*, London: Ward, Lock & Co., vol. XXII, June–November 1905, pp.643–52

p.15   **There were at least a dozen**: Edward Linley Sambourne, a renowned cartoonist and illustrator in the late nineteenth century, described a 'very funny' Sunday lunch that he had enjoyed with the Ingram family at The Bungalow on 18 December 1892. In his diary he said that a white jackdaw, presumably Darlie, was flying about the dining room, along with many other albino birds and a laughing jackass.

p.16   **John Jenner Weir, a friend**: 'My friends, in feather and fur' by Lady Ingram, pp.643–52

p.17   **'rushed out and brought back'**: Oral-history interviews by Anna Stirling Pope of her grandfather, Edward Stirling Booth, recorded at his home in Stirling, South Australia, on eighteen occasions between 1 February and 2 August 1995

p.18   **'The study of birds'**: *Random Thoughts on Bird Life* by Collingwood Ingram, self-published, 1978, p.1

p.19   **'I was only about 10'**: Ibid., p.2

2. *Mayfair-by-the-Sea*

p.22   **In the late 1860s**: *Wings Over the Western Front: The First World War Diaries of Collingwood Ingram*, edited by Ernest Pollard and Hazel Strouts, Charlbury, Oxfordshire: Day Books, 2014, p.1

p.23   **Sir Erasmus and Lady Charlotte Wilson**: Sir Erasmus Wilson was an eminent surgeon and dermatologist, who championed sanitary reform and frequent bathing. He was also known for paying £10,000 to transport Cleopatra's Needle, a 3,000-year-old granite obelisk, from Egypt to London, where it was erected in 1878 on the Thames Embankment.

p.23   **The bird was unable**: *The Migration of the Swallow* by Collingwood Ingram, London: H.F. & G. Witherby Ltd, 1974, p.21

p.23   **During the first decade**: *Random Thoughts on Bird Life* by Collingwood Ingram, self-published, 1978, p.1

3. *Triumphs and Tragedies*

p.24   **William Little and Nathaniel Cooke**: Nathaniel Cooke had become Herbert's brother-in-law in 1835 after marrying his sister, Harriet Ingram.

p.25   **'commenced by kissing me'**: *Herbert Ingram, Esq., M.P.* by Isobel Bailey, Boston: Richard Kay, 1996, pp.144–5, 158

p.27   **'sowed the seeds'**: *Isles of the Seven Seas* by Collingwood Ingram, London: Hutchinson & Co., 1936, p.144

p.28   **'absolutely identical'**: *The Birds of the Japanese Empire*, London: R.H. Porter, 1890, pp.36–7

p.28   **worked at the *Illustrated London News***: To amuse his wife, Emily, who was suffering from cancer, Wain started sketching their cat, Peter, wearing clothes and enjoying human hobbies. An illustration he had drawn in the *Illustrated London News* of a feline party became his springboard for a career drawing wide-eyed anthropomorphic comic cats and kittens playing poker, the piano and other pursuits. Ingram's artistic growth also undoubtedly benefited from discussions with the innumerable artists and illustrators who worked for his father's magazines and were frequent guests at their homes. These included Edward Linley Sambourne, who illustrated the 1885 edition of *The Water Babies*, and Sir William Quiller Orchardson, a Scottish portraitist and painter, who moved to Westgate-on-Sea in 1877.

p.28–9 **Percy Horace Gordon Powell-Cotton**: Powell-Cotton had also almost suffered a similar fate to Ingram's Uncle Walter.

In October 1906 he was badly mauled by a lion while on a honeymoon expedition with his wife, Hannah, on the banks of a river in the Congo Free State. Yet despite seventeen claw wounds, he escaped death, thanks to a folded copy of *Punch* magazine in his breast pocket, which prevented the lion from ripping open his abdomen. Porters shot the lion, which is on display at Quex Park museum. *Punch* immortalised the incident in its February 1907 edition:

The wounded lion with a lusty roar
Advanced to drink the gallant major's gore.
But suffered great confusion when he felt
An unexpected Punch below the belt.

p.30   **'a primordial instinct'**: *Isles of the Seven Seas*, pp.58, 66
p.30   **'Some Egyptian sparrows'**: 'My friends, in feather and fur' by Lady Ingram, pp.643–52
p.31   **'one of the most beautiful'**: *Isles of the Seven Seas*, p.272
p.31   **'rubbed their eyes'**: *Illustrated Sporting and Dramatic News*, 23 October 1897, pp.290–1
p.32   **'if I were not old enough'**: *Wings Over the Western Front: The First World War Diaries of Collingwood Ingram*, edited by Ernest Pollard and Hazel Strouts, Charlbury, Oxfordshire: Day Books, 2014, p.4

4. *Enforced Seclusion*

p.34   **The second period**: Strictly speaking, the *Sakoku* era lasted from 1639 to 1853.
p.36   **Vasco da Gama's voyage**: Among notable voyages of discovery and colonisation, the Portuguese captured the Indian city of Goa in 1510, the Malay city of Malacca in 1511 and set up a trading settlement in Ceylon (present-day Sri Lanka) in 1518. They reached China soon afterwards and started anchoring ships in Macau's harbour in the 1530s.
p.37   **And for more than two centuries**: Some Japanese scholars use the term *Kaikin*, or maritime provisions, rather than *Sakoku*, noting that Japan actively traded with the Dutch as well as with China and Korea. The period 1603–1868 is also known as the Edo period and the Tokugawa period.

5. *Japan Beckons*

p.43 **The Japanese people 'looked'**: *Recollections of a Happy Life*, by Marianne North, London: Macmillan and Co., 1892, vol.1, p.216

p.43 **'The stranger finds himself thinking'**: *Glimpses of Unfamiliar Japan*, by Lafcadio Hearn, New York: Cosimo Classics, 2005, p.21. Originally published by Berhard Tauchnitz in 1907.

p.43 ***The Mikado* at the Savoy Theatre:** Collingwood Ingram diary entry, 24 June 1896

p.44 ***The Geisha*, a musical comedy**: Ibid., 18 July 1896

6. *The Rising Sun*

p.46 **'I have never seen man'**: Collingwood Ingram undated diary entry, 1902

p.48 **'The war opened the eyes'**: *Japan in Transition* by Stafford Ransome, London: Harper & Brothers, 1899

p.49 **'The country seems steaming damp'**: Collingwood Ingram undated diary entry, September 1902

p.50 **'To see the hundreds of fans'**: Ibid., 9 September 1902

p.50 **'When the land was behind the horizon'**: Ibid., 20 September 1902

7. *The Birds and the Bees*

p.53 **The bird was named**: Ingram's quest to find the White's Thrush eggs originated after he met H.E. Dresser, a renowned British ornithologist, in London. Although bird experts described the White's Thrush as a British bird, no one had ever found its eggs. In Japan, the bird was occasionally seen in the wooded mountains. Dresser told Ingram that finding the White's Thrush's eggs, along with those of the Siberian Thrush, would give him 'oological prizes of the first order'.

p.53 **Permission obtained**: Twenty-four years later, in 1931, as Ingram was standing on the first tee of a golf course in Estoril, Portugal, another golfer suddenly said, 'Well, did you find your White's Thrush's nest?' The mystery questioner was Sir Francis Lindley, the British ambassador to Portugal at the time, whose job at the

UK Embassy in Japan in 1907 included helping Ingram get his bird-collecting permit. Lindley became British Ambassador to Japan later in 1931. *In Search of Birds* by Collingwood Ingram, London: H.F & G. Witherby Ltd, 1966, pp.26–7, 29

p.53 **'Grandpa did whatever he wanted'**: Collingwood's brother, Bertie, followed him to Japan, and in somewhat similar circumstances. Bertie had met Hilda Lake, daughter of a wealthy businessman, probably at Westgate-on-Sea's country club, according to Jackie Ingram, his granddaughter. Hilda, known to friends as Blossom, was sixteen years old and still at Roedean, an exclusive girls' boarding school, when they became engaged. He was about seventeen years older. The couple married in 1908 and sailed on honeymoon to Japan via Aden, Ceylon, Singapore and Hong Kong. Already interested in Japanese curios, Bertie started collecting *netsuke*, lacquered objects and Satsuma earthenware. These, together with Chinese greenware and ceramics, are today part of the Ingram Collection in the Ashmolean Museum at the University of Oxford.

p.55 **Sir William had heard**: *Isles of the Seven Seas*, p.218

p.55 **'in a state of helpless inebriation'**: Ibid., p.235

p.56 **The Thames estuary**: *Wings Over the Western Front*, ed. Ernest Pollard, p.8

p.56 **In 1914 the Royal Naval Air Service**: The Royal Naval Air Service and the Royal Flying Corps merged in April 1918 to become the Royal Air Force.

8. Ingram's War

p.57 **This depot serviced**: *Wings Over the Western Front*, ed. Ernest Pollard, p.11

p.57 **As well as raising her children**: I am grateful to Dr Dawn Crouch, an expert on Westgate-on-Sea, for her insight into the life of Sir William and his family in the town.

p.58 **'A moon of unsurpassable brilliance'**: *Wings Over the Western Front*, ed. Ernest Pollard, p.147

p.58 **'the uncared-for dead'**: Ibid., pp.61, 245. Another diary entry mentioned Ingram's scrutiny of the plane that belonged to Baron von Richthofen, a German flying ace officially credited with eighty air-combat victories. The baron had been shot down and killed near Amiens on 21 April 1918. The Red Baron's Fokker Dr.I 425/17 triplane was then taken to Le Crotoy aerodrome, which Ingram frequently visited. When Ingram arrived, seventeen days after von Richthofen's death, the officer in charge of salvaging the plane apologised that there was no blood left in the fuselage because 'the gore-bespattered pieces have been much sought after by the more eager and morbid souvenir hunters'. Ingram left with a small piece of bloodless fabric. Ibid., p.186.

p.59 **'The most galling evidence'**: Ibid., pp.62–3

p.59 **'The ravages of war'**: Ibid., p.223

9. *Birth of a Dream*

p.61 **'Ornithology has become'**: *Isles of the Seven Seas*, pp.142–3

p.63 **'a succession of sylvan glades'**: *A Garden of Memories* by Collingwood Ingram, London: H.F & G. Witherby Ltd, 1970, p.9

p.63 **'from what had been'**: Ibid., p.10

p.64 **'would be difficult to conceive'**: *Ornamental Cherries* by Collingwood Ingram, London: Country Life, 1948, p.23

PART TWO

10. *Twin Quests*

p.69 **'Because there is a certain similarity'**: *Isles of the Seven Seas*, p.147

p.70 **At that time**: A.E. Housman praised England's wild cherries in his poem of the 1890s:

> Loveliest of trees, the cherry now
> Is hung with bloom along the bough,
> And stands about the woodland ride
> Wearing white for Eastertide.

p.70 **Hence the reason**: *Japanese Flowering Cherries* by Wybe Kuitert, Portland, Oregon: Timber Press, 1999, p.75. Kuitert, a Dutch professor of environmental studies at Seoul National

University, is renowned in the horticultural world for this authoritative treatise on the blossoms.

p.71 **'A garden should be'**: *The English Flower Garden* by William Robinson, 1900 edition, pp.39, 159

p.72 **No other peoples**: In China, the cultivation of cherry blossoms is thought to date back to the Qin dynasty of 221–206 BC, when the flower was bred in royal gardens. *China Daily*, 30 March 2015; Chinadaily.com.cn

p.72 **All of these man-made varieties**: The cultivated varieties are also known as man-made, ornamental and flowering cherries. The ten known wild species are: Japanese mountain cherry (*Yama-zakura*), Ōshima cherry (*Ōshima-zakura*), Korean hill cherry (*Kasumi-zakura*), Fuji cherry (*Mame-zakura*), Sargent cherry (*Ōyama-zakura*), Japanese alpine cherry (*Takane-zakura*), Clove cherry (*Chōji-zakura*), Korean mountain cherry (*Miyama-zakura*), Taiwan cherry (*Kanhi-zakura*) and Spring cherry (*Edo-higan*). *Kanhi-zakura* grows in the wild only on an Okinawan island, and it is unclear whether it is native to the region or was introduced to the island from elsewhere. Hence it is sometimes excluded from Japan's 'native' species. In 2016 Toshio Katsuki, chief researcher at the National General Research Institute of Trees in Tokyo, discovered a new species of wild cherry on the Kii Peninsula in Japan. He named the early-blossoming pink-flowered cherry *Kumano-zakura*. For more on *Kumano-zakura*, see *Sakura no Kagaku* (*The Science of Sakura*) by Toshio Katsuki, Tokyo: SB Creative, 2018.

p.72 **Japan's well-defined seasons**: *Japanese Flowering Cherries* by Wybe Kuitert, pp.21–4.

p.73 **The first-known cultivated cherry**: *Sakura* by Toshio Katsuki, p.87

p.73 **In AD 812**: The link between the imperial family and cherries was made clear by the mid-ninth century AD, when a *Yama-zakura* tree was planted on the left-hand side of the main building of the Imperial Palace in Kyoto, replacing the traditional plum tree. Today the trees on either side of the palace are known as *Sakon-no-Sakura* (the cherry tree on the left) and *Ukon-no-Tachibana* (the *Citrus tachibana* on the right).

p.73 **At their annual *hanami* gatherings**: Japan's first collection of poetry, the *Man'yōshū*, a twenty-volume eighth-century master-piece, contained forty-three poems that cited cherries. A later collection, the *Kokin Wakashū*, published in about AD 905, included numerous cherry-blossom poems. In these, the poets interpreted the symbolism of the flowers in various ways. The most common themes were the fragility, futility and imperman-ence of life, love and beauty.

p.74 **In *The Tale of Genji***: *Kamikaze, Cherry Blossoms and Nation-alism: The Militarization of Aesthetics in Japanese History* by Emiko Ohnuki-Tierney, Chicago: University of Chicago Press, 2002, p.41

p.74 **'Let me die'**: During the twelfth century the samurai class emerged and started to threaten the established order in Kyoto. In 1192 Minamoto no Yoritomo established a shogunate in the eastern city of Kamakura. He appointed *daimyō* to rule the domains. The shogunate system, also known as the *Bakufu* system, essentially governed the nation for the following seven centuries, until the Meiji Restoration. The emperor became a remote figure whose main role was to perform ancient rituals. As power migrated from west to east, and as the *daimyō* and samurai started travelling back and forth, cherry trees became a symbol of the transition.

This is how the Ōshima cherry ended up in Kyoto. A species with large white flowers, it grew naturally in the east, in the Izu and Bōsō peninsulas near Kamakura. Since the shōgun also maintained offices in Kyoto to oversee western Japan, the cherry accompanied his entourage. Two other cherries developed in Kamakura also migrated to the west: *Fugenzō* and *Mikuruma-gaeshi*.

p.75 **This system meant that**: Many high-ranking *daimyō* had three residences in Edo alone. One was located next to Edo Castle, where the shōgun lived with his official wife and family; it func-tioned almost like an embassy. Inside this home, the residents had extraterritorial rights that were outside the shōgun's control. A second residence was for retired *daimyō* or grown-up heirs. A third was usually a more leisurely home with a large garden, further away from the centre of Edo. In all, these *daimyō* resi-dences occupied more than one-third of all the land in the city.

p.75 **After years of repeated trial**: *Sakura* by Toshio Katsuki, p.93

p.75 **This grafting technology**: Fujiwara no Teika, an influential aristocratic scholar and poet in the early thirteenth century, wrote in his diary, the *Meigetsuki*, that he had grafted a cherry tree in his garden, according to *A Literary History of Cherries* by Kazusuke Ogawa, Tokyo: Bungeishunjū, 2004, p.72

p.76 **While there is no accurate record**: *Nihon no Sakura* by Minoru Okuda, Hiroshi Kihara and Tetsuya Kawasaki, Tokyo: Yama-kei Publishers, 1993, p.4

p.76 **'branches so gentle'**: *Japan Times*, 25 March 2012

p.76 **Another high-ranking *daimyō***: *Japanese Flowering Cherries* by Wybe Kuitert, p.15

p.77 **Several other *hanami* spots**: In the early eighteenth century the eighth shōgun, Yoshimune Tokugawa, ordered thousands of cherry trees to be planted in three places: the east bank of Edo's Sumida River, the banks of the Tama aqueduct in Koganei and in Asukayama. At the time it was believed that the roots of cherry trees purified water underground. The trees also helped to strengthen the riverbanks and became prime *hanami* spots for city dwellers.

11. *The Dejima Doctors*

p.80 **As Ingram noted**: *Ornamental Cherries*, p.62

p.80 **'[At] the beginning of the spring'**: *The History of Japan* by Engelbert Kaempfer, published posthumously in 1727, Glasgow: James MacLehose and Sons, 1906 edition, vol. 2, p.24

p.80 **'to prevent any thoughts'**: Ibid., p.325

p.84 **There he published widely**: The materials taken by Siebold formed the basis of the National Museum of Ethnology in Leiden, a town in the southern Netherlands. Books and journals by, and about, the Dejima doctors perpetuated their legacy, as did numerous plants: a pale-pink, semi-double flowering cherry called *Prunus sieboldii*, also known as *Takasago*; a black pine tree named *Pinus thunbergii*; and the genus *Caulokaempferia* K. Larsen (Zingiberaceae).

p.84 **In the rarefied world**: In March 1877 Gon Abe played a small role in Japan's last civil war. He joined hundreds of volunteer doctors and nurses from across Japan to treat the dying and injured at the

Battle of Tabaru Slope in Kumamoto, a town in Kyushu about 120 miles from Nagasaki. The doctors and nurses were part of the Philanthropic Society, the forerunner of the Japanese Red Cross Society. There was no shortage of work. Thousands died or were wounded when as many as 90,000 heavily armed government troops fought about 15,000 former samurai who were upset with the Meiji government's reforms. Leading the former samurai forces was Takamori Saigō, from Kagoshima in southern Kyushu. Once a hero of the forces that overthrew the shogunate, he later clashed with the new government and became leader of the frustrated former samurai. On 25 September 1877, with defeat staring the rebels in the face, one of Saigō's loyal followers lopped off his leader's head with a single sword stroke to preserve his dignity, and the forty remaining rebels charged down the slope to their deaths. The armed uprising against the Meiji government was over.

After the battle, Gon Abe travelled more than 1,000 miles from Kumamoto to Aomori, in the far north of the island of Honshu, where he opened a clinic. Accompanying him were his wife, Yoshino, and their nine-year-old son, Hyakutarō. Both Hyakutarō and Hyakutarō's own son, Takatomo – the grandfather I never met – also became doctors, the last of at least fourteen generations of Abes in the same profession.

12. *Hunting Plants*

p.85   **Among Fortune's favourite exports**: Many Japanese botanists believe that the first Japanese flowering cherry tree that came to England was *Albo Plena* (*Prunus serrulata* 'Albo Plena') in 1822. It was brought from Canton by a British plant-hunter, Joseph Poole. For a long time the tree was thought to be Chinese. But some experts claim this tree is the same as a Japanese variety called *Ichihara-tora-no-o*, or tiger's tail. They believe the *Albo Plena* originated in Japan and came to England via China.

p.85   **'All countries are beautiful'**: *Yedo and Peking* by Robert Fortune, London, John Murray, 1863, p.183

p.85   **Sargent's interest was**: Sargent published *Forest Flora of Japan* in 1894, based on his 1892 visit. He called the cherry the largest tree of the rose family, more cultivated for flowers in Japan than any tree apart from the apricot. 'In the early autumn it is conspicuous in the landscape and very beautiful, as the leaves turn deep scarlet and light up the forest before the maples assume their brightest colours,' he wrote.

p.85 **the botanical artist Marianne North**: Many of Marianne North's paintings are on display in the Marianne North Gallery at Kew Gardens.

p.86 **'Why should the trees'**: *Glimpses of an Unfamiliar Japan* by Lafcadio Hearn, Boston: Houghton Mifflin, 1894. There was also a British writer, Reginald J. Farrer, who had much in common with Collingwood Ingram, being a self-taught naturalist who did not attend school, in this case because he had a speech defect. Born a few months earlier than Ingram, in February 1880, Farrer had graduated from Oxford University in 1902 and set off for China, Korea and Japan to explore their botanical heritage. The differences between flower-lovers in Britain and Japan were profound, he wrote:

A flower, to be admitted by Japanese canons, must conform to certain rigid rules. At the head of rejected blossoms stand the rose and the lily, both of whom are considered by the Japanese rather crude, unrefined efforts of nature. The elect are cherry, wisteria, peony, willow-flower, iris, magnolia, azalea, lotus, peach, plum and morning glory. This is the hierarchy. And for his favourites, no attention is too onerous. (From *The Garden of Asia – Impressions from Japan* by Reginald J. Farrer. London: Methuen & Co., 1904, p.21.)

p.86 **'Old Japan is dead and gone'**: *Things Japanese* by Basil Hall Chamberlain, London: John Murray, 5th revised edition, 1905, p.16

p.87 **'If someone asked'**: The original Japanese was:

Shikishima no Yamato gokoro o
Hito towaba
Asahi ni niou
Yama-zakura bana.

p.87 **'The cherry is first among flowers'**: The original Japanese was:

Hana wa sakuragi,
hito wa bushi.

p.87   **Nitobe wanted to help**: The life and philosophy of Nitobe, who became Under-Secretary General of the newly founded League of Nations in 1920, is captured in the exquisite Nitobe Memorial Garden at the University of British Columbia, outside Vancouver in Canada.

p.87   **'spirit of *bushidō'***: *Bushidō: The Soul of Japan* by Inazō Nitobe, 13th edition, 1908, Project Gutenberg e-book

p.88   **His book, first published in 1871**: Mitford's political contemporary and friend in Japan was Ernest Satow, who lived there from 1862 to 1883. In 1881, Satow planted a cherry tree in front of the British legation building, which triggered a mass planting of cherry trees along the moat of the Imperial Palace by the Tokyo City government 17 years later. Today, the cherry avenue in front of the British Embassy and along the Imperial Palace moat is one of the most famous cherry-viewing spots in Japan. Satow returned as Envoy Extraordinary in 1895 for five years.

p.88   **'The feudal system has passed away'**: *Tales of Old Japan* by Algernon Freeman-Mitford, London: Macmillan and Co., 11th edition, 1910, Preface, p.viii

p.89   **More than eight million people**: An entrance gate and an avenue of stone lanterns were installed at the White City site in August 2018, to create a cherry tree-lined path leading visitors towards the Japanese gardens.

p.90   **Fairchild and his wife**: *The Cherry Blossom Festival: Sakura Celebration* by Ann McClellan, Boston: Bunker Hill Publishing, 2005, pp.28–9. See also *The World Was My Garden: Travels of a Plant Explorer* by David Fairchild, New York: Charles Scribner's Sons, 1938. Besides cherries, Fairchild was responsible for introducing many plants and crops into the US, including soya beans, mangoes and pistachios.

p.91   **Undaunted, Mayor Ozaki**: The cherry-blossom varieties sent to the US in 1912 included *Ariake, Fugenzō, Fukurokuju, Gyoikō, Ichiyō, Jō-nioi, Kanzan, Mikuruma-gaeshi, Somei-yoshino* and *Taki-nioi*. About 1,800 of the 3,020 trees sent to Washington were *Somei-yoshino*. Twenty *Gyoikō* trees were planted in the grounds of the White House.

p.91   **After arriving in Seattle**: In May 2011, New York City officials and Japan's consul-general planted cherry trees in Central

Park in memory of the victims of the Tōhoku earthquake and tsunami.

p.91 **Scidmore watched President Taft's wife**: Eliza Scidmore, whose letter prompted the export of cherries to the US, died in Geneva in 1928. She is buried in the Yokohama Foreigners' Cemetery, where she lies next to her brother, a diplomat in Japan, and her mother, in the shade of a cherry tree grafted from a plant on the Potomac.

p.91 **In this endeavour**: In the early 1920s Ingram had tried his hand as a vintner for a few years. He always enjoyed his wine, saying that the wine god Bacchus was indeed 'a benevolent deity'. As a wine merchant, he had a convenient excuse to visit Portugal and Spain in 1922 and 1923 to collect plants and view birds. *Isles of the Seven Seas*, p.50

13. *Creation and Collection*

p.93 **'My aim was to reduce'**: *A Garden of Memories*, p.9
p.93 **During the 1900s Bean**: Article by Tony Kirkham in *The English Garden*
p.94 **Still more came from Clarence McKenzie Lewis**: Lewis's mother, Helen Forbes Lewis Salomon, was the widow of William Salomon, founder of Salomon Brothers, the New York investment bank. Lewis's wife, Annah, a rose-lover, died in 1918. His mother was widowed in 1919. The state of New Jersey bought Skylands in 1966 and it is now the state's official botanic garden.

14. *The Hokusai Connection*

p.95 **The sweet-smelling, light-pink blossoms**: After the garden broadcaster Marion Cran learned from Ingram that *Amanogawa* means the 'Milky Way' in Japanese, she wrote: 'The Amanogawa is a streak of white loveliness in the garden, a long slender line of whiteness (like the Milky Way). Why don't we use those pretty words for the cherries?' *Queen* magazine, 30 November 1932, p.31

p.96 **It was also called**: *Japanese Flowering Cherries* by Wybe Kuitert, pp.281–2 p.96 **'In attempting to create'**: 'Notes on Japanese Cherries', part II, *Journal of the Royal Horticultural Society*, vol. 54, 1929, p.161

p.97 **As great rivals**: *Japanese Flowering Cherries* by Wybe Kuitert, p.89

p.97 **'I do not suggest'**: 'Notes on Japanese Cherries', part I, *Journal of the RHS*, 1925, p.74

p.97 **'the world's most shapely mountain'**: 'On one unforgettable occasion, I saw its awe-inspiring splendour,' Ingram wrote about his visit to Mount Fuji in 1926. 'While the base of the great volcano was still plunged in nocturnal darkness, its summit, draped in a tattered shawl of snow, was already aglow with the rosy rays of an unseen sun. Nowhere have I been so deeply impressed as I was that morning by the breathtaking beauty of Fujiyama, its snowy summit pink-flushed and glorified by the rays of a yet-unseen sun.' *In Search of Birds*, pp.35–6

p.98 **Ingram, himself an accomplished artist**: Hokusai famously wrote that until he was seventy, 'nothing I drew was worthy of notice. At 73 years, I was somehow able to fathom the growth of plants and trees, and the structure of birds, animals, insects and fish.'

p.98 **'When, in the spring'**: *Ornamental Cherries*, p.232

p.99 **A wealthy ornithologist**: Duke Takatsukasa's son, Toshimichi, married Emperor Hirohito's daughter, Princess Kazuko. In 1966, Toshimichi and his mistress, a Ginza nightclub hostess, were found dead of carbon monoxide poisoning.

p.99 **Takatsukasa was particularly impressed**: Duke Takatsukasa, along with a fellow student Count Nagamichi Kuroda, had founded the Japan Ornithological Society in 1912 and had installed Professor Iijima as its first president. After Iijima's death in 1922, the duke, then in his early thirties, became the society's head. Following his Benenden visit, Takatsukasa not only invited Ingram to visit Japan, but also became honorary president of the Cherry Association (*Sakura No Kai*), the group of cherry-tree enthusiasts in Tokyo whose members were the ruling elite. In a society where personal connections were of the highest priority, the duke's friendship with Ingram opened doors that few foreigners could even imagine.

p.114 **'An apple orchard in the Alps'**: Collingwood Ingram diary entry, 16 April 1926

p.116 **The best viewing spot**: Years earlier, in December 1910, Funatsu and some government workers selected and cut the Arakawa River cherry scions that were grafted and later sent as saplings to Washington D.C. and New York by Tokyo's Mayor Ozaki.

p.116 **'One could only marvel'**: Miyoshi in *Kōhoku no Goshiki-zakura*, Tokyo: Kōhoku History Society, 2008, pp.153–4

p.116 **Funatsu then helped Miyoshi**: 'Everything else paled in comparison to the pleasure of those days when [Funatsu and I] were able to examine such numerous cherries together,' Miyoshi wrote in *Kōhoku no Goshiki-zakura*, p.154

p.117 **'The love-light in his eyes'**: 'The cult of the flowering cherry in Japan', by Collingwood Ingram, *The Garden Chronicle*, 20 November 1926

p.118 **'decaying or on the point of death'**: Collingwood Ingram diary entry, 20 April 1926

18. *Guardian of the Cherries*

p.119 **Magoemon's pursuit**: *Kōhoku no Goshiki-zakura*, 2015, p.62

p.120 **Villagers planted them**: As well as being mayor, Shimizu operated a private school, or *juku*, at his Kōhoku home, where he taught Chinese classics and mathematics. Seisaku Funatsu had been one of Shimizu's best students a few years earlier.

p.121 **Without his enthusiasm**: 'The Cult of the Cherry Blossom', *Illustrated London News*, 28 April 1934, p.610

p.121 **Many people considered**: Interview with Tetsu Tada, curator of the Koganei Cultural Property Centre, 22 December 2017

p.122 **Desperate to save the tree**: Collingwood Ingram diary entry, 21 April 1926

p.123 **He asked its wealthy owner**: Isomura's cherry scions arrived in Benenden in the winter of 1926, along with the scions of some red plum trees.

19. *Wild-Cherry Hunting*

p.123 **Better still, when Ingram**: Collingwood Ingram diary entry, 26 April 1926

p.124 **'were growing on a distant slope'**: *Ornamental Cherries*, p.87
p.124 **'Viewed against the dazzling blue'**: Ibid., p.88
p.124 **'Here and there'**: Collingwood Ingram diary entry, 6 May 1926
p.125 **'liquid pipe of the *uguisu'***: *Ornamental Cherries*, p.88. Nitobe's popular book on *bushidō* noted that samurai were encouraged to express their gentler emotions in verse. One poem that he quoted bore a similarity to Ingram's poignant diary notes about the *uguisu*, or nightingale:

> Stands the warrior, mailed and strong
> To hear the *uguisu*'s song
> Warbled sweet the trees among.

p.125 **'in its full vernal glory'**: *A Garden of Memories*, pp.62–4
p.126 **'Every twig was closely packed'**: Collingwood Ingram diary entry, 29 April 1926

20. *Saving the* Sakura

p.128 **Additionally he believed**: *Sakuramori* was also the name of a best-selling novel published in 1976 by Tsutomu Mizukami, about a man who devoted his life to preserving cherry trees.
p.129 **'make known to the world'**: *Sakura* journal, April 1918, no. 1, pp.1–2
p.130 **The honorary president was**: After Marquis Yorimichi Tokugawa's death in 1925, Duke Takatsukasa became the Cherry Association's honorary president.
p.131 **The final straw**: *120 Years at the Imperial Hotel*, Tokyo: Imperial Hotel Ltd, 2010, p.28
p.131 **'maintain a pretext'**: *Sakura* journal, April 1918, no. 1, pp.1–2
p.132 **'No easy task is before us'**: *Sakura* journal, April 1919, no. 2, pp.2–4
p.133 **'The cherry blossom is essentially'**: *Sakura* journal, April 1919, no. 2, pp.4–6

21. *Ingram's Warning*

p.134 **'Many cherry trees in Japan'**: *Sakura* journal, April 1927, no. 9, pp.5–6

When Ingram began to speak: Collingwood Ingram original speech notes, April 1926

PART FOUR

22. *The Restoration Quest*

p.139 **'This was a pity'**: 'Notes on Japanese Cherries', part II, *Journal of the RHS*, 1929, p.162

p.141 **They needed little supervision**: Florence had her own life at The Grange. Among other plants, she grew sweet peas in her own patch of garden and kept hens in the paddock.

p.141 **'refined charm'**: *Isles of the Seven Seas*, p.260

p.142 **During the winter of 1926**: The 1927 edition of the Cherry Association's *Sakura* journal contained a mention of the cuttings sent to Ingram earlier that year.

p.142 **Opening each crate**: *A Garden of Memories*, p.69

p.143 **Once they had sprouted**: 'Some Plant-Hunting Experiences' by Collingwood Ingram, *Journal of the RHS*, vol. LXXXI, part 10, October 1956, p.440

p.143 **Edward Augustus Bowles**: Bowles was the great-uncle of Andrew Parker Bowles, whose first wife, Camilla Shand, became Duchess of Cornwall in 2005 after marrying Prince Charles.

p.143 **His methods required**: *Ornamental Cherries*, pp.29, 32–4

p.144 **He had hoped**: Letters from Paul Russell, assistant botanist at the US Department of Agriculture's Bureau of Plant Industry, 12 February 1930 and 28 January 1931. Russell's 1930 letter said that scions of *Taihaku*, which Ingram sent in summer 1928, did not survive.

p.144 **For friends outside England**: *Ornamental Cherries*, p.31

23. Taihaku's *Homecoming*

p.144 **he visited the Greyfriars Estate**: The Greyfriars Estate, named after the Franciscan friary in its grounds, later became a special-needs home. See *Hunting Down the Great White Cherry* by Owen Johnson and Ernest Pollard, *The Garden*, date unknown

p.145 **In 1899**: *Ornamental Cherries*, p.207

p.145   **'rarity and remarkable beauty'**: Ibid., p.208

p.146–47 **'Piqued by his obvious doubt'**: Ibid., p.209

p.147   **'withered away'**: *Sakura* journal, 1927, no. 9

p.147   **Founded in AD 888**: In the preface to his Kyoto cherry book, one of five that he penned between 1931 and 1940, Kayama wrote: 'Since I was a little boy, I felt inexpressible nostalgia and affection towards *sakura*, perhaps because I grew up in Omuro, which is renowned for cherries. That passion grew even stronger as an adult. Every spring, whenever my work schedule allowed, I travelled all over Kyoto in search of cherries.' In 1943 Kayama and his son, Tokihiko, co-wrote another book on cherries.

p.148   **Besides protecting the trees**: In 1961 the fifteenth Tōemon Sano published a book called *Sakura: Flowering Cherries of Japan*, which lists 101 cherry species and varieties.

p.151   **To Sano, *Taihaku***: In 2013 a Japanese research group established, from a survey of DNA, that the cherry varieties of *Komatsunagi*, *Kurumadome* and *Taihaku* were all the same. Since then it has been speculated that *Taihaku* might still exist in Japan under a different name. However, Toshio Katsuki, head researcher of the Forestry and Forest Products Research Institute (FFPRI), believes there is a high probability that *Komatsunagi* and *Kurumadome* scions were mistakenly swapped with those of *Taihaku*, after its return to Japan from England. In Ingram's 1926 journal, one of the cherry trees seen on the banks of the Arakawa River was described as being *Komatsunagi* and as having different characteristics from *Taihaku*.

p.153   **'I feel it might be a disgrace'**: *Sakura Otoko Gyōjō* by Shintarō Sasabe, Tokyo: Heibon-sha, 1958, pp.283–4.

p.153   **He also stated**: Mizukami's historical fiction novel, *Sakuramori*, includes the incident.

p.154   **Sad to say**: In 2016, after reading the Japanese edition of this book, a cherry researcher called Keiichi Higuchi visited the sixteenth Tōemon Sano in Kyoto to see the third-generation *Taihaku*. Sano gave Higuchi a branch from the tree, which Higuchi kept in a vase at his Tokyo home. When it flowered in spring 2017, he and Funatsu's 89-year-old granddaughter-in-law, Setsuko, placed the flowering branch on the Funatsu family's grave near the Arakawa River.

p.154 **'From that tiny nucleus'**: *Ornamental Cherries*, p.208. Today, *Taihaku* is far more popular in Britain, the US and Australia than it is in Japan, where the *Somei-yoshino* variety rules supreme. *Taihaku* thrive in Tokyo's Shinjuku Gyoen Park, once the residence of a *daimyō* lord and now a national park. They can also be seen in research-focused centres, such as Hachiōji city's Tama Forest Science Garden near Tokyo and Mishima city's National Institute of Genetics in central Japan.

p.155 **An earlier letter from Kayama**: It is unknown when Kayama and Ingram stopped corresponding, or whether Ingram eventually learned of Kayama's death on 5 November 1944, ten months before the end of the Second World War. The head of his neighbourhood association, Kayama contracted pneumonia while delivering rationed food to Kyoto residents in the rain. He was fifty-eight. Email interviews with Kayama's grandson, Yukihiko Kayama, retired Professor of Cerebral Physiology at Fukushima Medical University, in May 2018

p.155 **'Oh cherries, cherries'**: Quoted in 'The Cherries of Omuro' by Collingwood Ingram, *Gardening Illustrated*, 9 April 1932

24. *Gambling with Success*

p.156 **Since then, Umineko cherries**: In 1970 Albert Doorenbos raised an almost identical tree to *Umineko* in the Netherlands and dubbed it *Snow Goose*.

p.157 **No one had ever deliberately hybridised**: Ingram was the first person in the world to try and hybridise cherries, according to Hiroyuki Iketani, a senior researcher at the National Agriculture and Food Research Organization.

p.157 **'Hybridising plants'**: *Gardening Illustrated* magazine, July 1952, p.184

p.158 **'The captain would kneel'**: Interview with Ruth Tolhurst, October 2014

p.159 **'In short, what I had'**: *Gardening Illustrated*, July 1952, p.185

p.159 **'of my all-too-often futile attempts'**: *A Garden of Memories*, p.11

p.160 **'be appreciated by all'**: 'Breeding New Flowering Cherries', *Gardening Illustrated*, July 1952, p.184

p.161	**But with Ingram's midwifery**: Ingram thought at first, erroneously, that the parents were the Kurile and Sargent cherries, hence the name Kur-Sar. *Japanese Flowering Cherries* by Wybe Kuitert, p.168
p.161	**In all, he created**: RHS's International Rhododendron Register, vol. 2

25. *A Fairy-Tale Garden*

p.161	**'When you entered the gates'**: Interview with Patricia Thoburn, October 2014. Thoburn died in October 2017.
p.162	**And a handsome gorse-like broom**: *Journal of the RHS*, vol. LXXXI, part X, October 1956, p.439
p.162	**'some dimly remembered scene'**: *A Garden of Memories*, p.11
p.163	**'When a new tree flowered'**: Interview with Ruth Tolhurst, October 2014
p.164	**'I had to climb in my van'**: Interview with Peter Kellett, October 2014
p.164	**It's a 'sturdy, stocky little tree'**: 'Notes on Japanese Cherries', part II, *Journal of the RHS*, vol. 54, 1929, pp.176–7
p.165	**In 1925 Ingram had forty**: In a 1925 paper for the *Journal of the Royal Horticultural Society*, called 'Notes on Japanese Cherries', Ingram wrote about both wild and cultivated cherries, including the pink-blossomed *Shujaku* variety, named after a mythical firebird, and the reddish-pink double-flowered *Kirin*, named after a mythical hoofed beast called the *kylin* or *qilin*. Uncertain of the heritage of twenty-nine varieties, he didn't mention them at the time, although he fastidiously recorded the details of new trees in the garden and sketched the structure, colour and shape of their flowers. Among the fifty-nine wild and cultivated cherries that he classified in 1929 were *Asano*, which he had discovered on the 1926 cherry hunt, and *Imose* and *Taoyame*, which he had found in Kyoto and introduced to Britain for the first time.
p.165	**'doubtful distinction'**: 'Notes on Japanese Cherries', part II, *Journal of the RHS*, vol. 54, 1929, p.179

26. *'Obscene'* Kanzan

p.166 **'are far too artistically minded'**: 'Some Plant-Hunting Experiences', *Journal of the RHS*, October 1956, p.443

p.166 **'It flaunts its finery'**: *Ornamental Cherries*, p.24

p.166 **One spring, Patricia Thoburn recalled**: Interview with Patricia Thoburn, October 2014

p.167 **'an inexcusable violation'**: 'Flowering Cherries in England' article, unknown US magazine, 1950

p.167 **'They're prostitutes'**: Interview with Peter Kellett, October 2014

27. *The Cherry Evangelist*

p.168 **'Every gardener is a plant hunter'**: 'Some Plant-Hunting Experiences', lecture to the RHS by Collingwood Ingram, 17 July 1956

p.169 **The club, called the Garden Society**: *The Garden Society 1920*, a short history of the society by Sir Giles Loder and others, was published in 1932. An expanded edition of the pamphlet was issued in 1996.

p.169 **Ingram's other Garden Society friends**: Other Garden Society members included Charles G.A. Nix of Tilgate Forest Lodge in Crawley, an authority on apples and pears; William Rickatson Dykes, secretary of the Royal Horticultural Society and an expert in breeding irises; Sir Frederick Stern, who created Highdown Gardens out of a Sussex chalk pit; and Henry Duncan McLaren, the 2nd Baron Aberconway, who developed Bodnant Garden in Conwy, Wales, and specialised in growing rhododendrons. Additionally, H.J. Elwes, a lily- and iris-lover, who developed the Colesbourne Estate in the heart of the Cotswolds; Geoffrey Taylour, the 4th Marquess of Headfort from County Meath in Ireland; Mark Fenwick from Abbotswood in the Cotswolds; Sir John Stirling-Maxwell, a Scottish politician and philanthropist; Sir William Lawrence, the founder of the Alpine Garden Society; and Edward Augustus Bowles, who developed the gardens at Myddelton House near Enfield.

p.170 **In sum, the Garden Society**: The Garden Society was just one of the British associations that focused on the natural world. Others included the British Ecological Society (founded 1913), the British Empire Naturalists' Association (1905), the Zoological Society of London (1826) and the Linnean Society of London (1788). Ingram was elected a fellow of the Linnean Society in 1944.

p.170 **Vita Sackville-West**: In a column in *The Observer* on 14 December 1952 Vita Sackville-West wrote about the winter-flowering *Fudan-zakura* cherry that Ingram liked. 'I don't like recommending plants of which I have no personal experience,' she wrote. 'But the advice of Captain Collingwood Ingram, the "Cherry" Ingram of Japanese cherry fame, is good enough for me and should be good enough for anybody.'

p.170 **'scraps of knowledge'**: 'On the Flowering Cherries' by Marion Cran, *The Queen* magazine, 30 November 1932, p.31

p.170 **The money raised**: William Rathbone, a Liverpool merchant and philanthropist who was friendly with Florence Nightingale, helped to establish a district nurses' movement in the city in the 1860s to care for the poor. The movement spread nationwide, with the support of Nightingale and Queen Victoria.

p.171 **'I get the impression'**: Interview with Nick Dunn, April 2015

p.171 **including Notcutts nursery**: Notcutts' founder, Roger Crompton Notcutt, and his eldest son, Roger Fielding (Tom) Notcutt, both became cherry aficionados because of Ingram and called him 'one of the greatest cherry authorities' in the world. See R.C. Notcutt and R.F. Notcutt, 'Flowering Cherries', *Journal of the RHS*, 1935, p.354

p.172 **'And there is absolutely'**: *Illustrated London News*, 28 April 1934, p.641

p.172 **In all, Ingram introduced**: *A Garden of Memories*, p.202

p.172 **Between 1928 and 1966**: Ingram's prize list was compiled by Brian Young, Ernest Pollard's brother-in-law. Pollard is Ingram's grandson-in-law.

28. Darwin Versus the Church

p.174 **Before the war**: Information about Benenden's history from Ernest Pollard's website: http://www.benenden.history.pollardweb.com/

p.175 **'My father and Ingram'**: Ingram despised religious hypocrisy. On a visit to Coll, an island in the Outer Hebrides, he had met an overweight minister of religion with whom he tried to discuss the finer points of theology. The preacher had no interest in listening to secular theories, and Ingram lambasted him as a 'narrow-minded and bigoted cleric. Despite his fanatical dogma, his stomach was his first and greatest God.' *Isles of the Seven Seas*, p.86

p.175 **'not entirely unlike'**: *Isles of the Seven Seas*, pp.75–6

p.177 **'war with Germany'**: *In Search of Birds*, p.143

29. *The Sounds of War*

p.177 **'The cherry blossoms danced'**: Interview with Patricia Thoburn, October 2014

p.179 **'From every side'**: *In Search of Birds*, p.251

p.179 **'Last night was one'**: Collingwood Ingram diary entry, 7 July 1917

p.179 **In particular**: *Random Thoughts*, p.3

p.180 **'the sound waves'**: *In Search of Birds*, p.251

p.182 **The sound would sustain**: The song 'A Nightingale Sang in Berkeley Square' was first performed in London in the *New Faces Revue* in April 1940. It became a hit later that year for Vera Lynn in the UK and for Glenn Miller and his orchestra in the US.

PART FIVE

30. *Cherry Blossom Brothers*

p.186 **History lessons began**: According to the *Kojiki* (*Records of Ancient Matters*), the sun goddess Amaterasu sent her grandson, Ninigi-no-Mikoto, from the sky to govern Japan. Ninigi met and married a beautiful goddess called Konohana Sakuya Hime, a 'blossom princess'. Some folklore experts say that Konohana Sakuya Hime represented the *sakura*, or cherry blossom. Her grandson, Jinmu, was said to be the first Japanese emperor in 660 BC, although there is little evidence that such a person actually existed.

p.189 **The infantry soldiers' collar**: The Myanmar armed forces still sing Burmese lyrics to the tune of *Hohei no Uta*. It's one of several

Japanese melodies that are popular in Myanmar, dating back to the Pacific War, according to Aung San Suu Kyi. *Letters from Burma* by Aung San Suu Kyi, London: Penguin Books, 1995, p.91

p.191 **Yama-zakura, Yama-zakura**: *Nejimagerareta sakura* (*Distorted Cherry Blossoms*) by Emiko Ohnuki-Tierney, Tokyo: Iwanami Shoten, 2003, p.212

p.192 **'Japan is a small country'**: Ibid., p.204

31. *Flowers of Mass Destruction*

p.194 **'the most artistic'**: 'Notes on Japanese Cherries', part II, *Journal of the RHS*, 1929

p.194 **'sooty ugliness'**: *Isles of the Seven Seas*, p.149

p.195 **Never before had**: The phrase 'flowers of mass destruction' is from *Flowers That Kill: Communicative Opacity in Political Spaces* by Emiko Ohnuki-Tierney, Redwood City: Stanford University Press, 2015, p.13

p.195 **As the emphasis**: *Kamikaze, Cherry Blossoms and Nationalisms* by Emiko Ohnuki-Tierney, pp.281–4

32. *Emperor Worship*

p.197 **One cause of indignation**: The unjustness of the treaties became apparent when a British cargo ship sank in the sea off Wakayama, east of Osaka, in October 1886. The ship's captain and twenty-five British and German crewmen escaped in lifeboats, but twenty-five Japanese passengers were left to drown. The Japanese government was hindered from investigating the case because of the treaty, and the captain, John Drake, was declared innocent of any wrongdoing by the British consulate in Kobe. Eventually, because of Japanese uproar, the case was tried at the British court in Yokohama. While the captain was sentenced to three months imprisonment, no compensation was paid to the victims' families.

p.197 **Another was that Japan**: Japan sought for decades to amend the unequal treaties of 1858 that the shogunate had signed. They were finally fully amended in 1911, after which Japan was allowed to exchange ambassadors with Western countries. The West did not recognise Japan as a first-class nation until after the Russo-Japanese War in 1904–5.

p.197 **In 1882 a Japanese delegation**: In Vienna a conservative German economist, Lorenz von Stein, drew the Japanese delegates a human body and suggested that his visitors use it as the model for the nation-state. The head was the monarch; the shoulders were the two parliamentary bodies; the arms were the army and the navy; the torso represented government institutions. A nation is only healthy, von Stein said, when its different parts are coordinated.

p.198 **Itō concluded**: Itō, who studied at University College London, was prime minister four times. He was assassinated in 1909 in Manchuria by a Korean independence activist.

p.198 **Together, these eventually formed**: After the Second World War, Japanese academics trying to understand the psychological, political, military and social influences that had led Japan to war called it the *tennō-sei*, or emperor-centred, ideology.

p.199 **In essence, the emperor**: *The People's Emperor* by Kenneth J. Ruoff, Cambridge, MA, Harvard University Asia Center, 2002, p.18

p.199 **The new structure**: Article 1 of the Constitution stated: 'The empire of Japan shall be reigned over and governed by a line of emperors unbroken for ages eternal.' Article 3 added: 'The emperor is sacred and inviolable.' Other articles defined the emperor as the 'sovereign head of the empire' and the supreme commander of the army and navy.

p.200 **'The Constitution had'**: Naoko Abe interview with Professor Kiyoko Takeda in *Sekai* magazine, October 2017. Takeda died in April 2018.

p.200 **In the carriage behind**: Two influential political scholars – Sakuzō Yoshino and Tatsukichi Minobe – wrote that a constitutional monarchy and democracy were compatible. In 1912 Minobe published the 'Emperor Organ' theory, arguing that the emperor was the organ of the state rather than a sacred power. This liberal interpretation was attacked in the 1930s.

p.201 **Bushidō's values**: Before the 1880s the word *bushidō* was rarely used in Japan, according to Oleg Benesch, not least because when the samurai class was abolished in the 1870s, many former warriors became so impoverished that the idea of a samurai-based ethic had little appeal. That changed as a new and confident generation came to power, particularly after Japan's victory against China in 1895 in the Sino-Japanese War. For more on this, see *Inventing the Way of the Samurai* by Oleg Benesch, Oxford University Press, 2014, pp.11, 15.

p.201 **Inazō Nitobe**: While Nitobe was a liberal Christian internationalist, another man promoted *bushidō* for starkly different purposes. Tetsujirō Inoue, a conservative anti-Christian nationalist, linked *bushidō* to Shinto creation myths and the emperor system, as a means of convincing the Japanese to accept the emperor's authority.

p.201 **They applied to everyone**: Another influential book, *Hagakure, the Book of the Samurai*, also spread the gospel of *bushidō*. Written in the early eighteenth century by Tsunetomo Yamamoto, *Hagakure* was not published until 1906, when *bushidō* was in vogue.

p.201 **By the late nineteenth century**: *Nejimagerareta sakura* by Emiko Ohnuki-Tierney, p.193

33. *The* Sakura *Ideology*

p.201 **'ever ready to depart life'**: *Bushidō: The Soul of Japan* by Inazō Nitobe, 13th edition, 1908, Project Gutenberg e-book

p.202 **In the late nineteenth century**: *Nejimagerareta sakura* by Emiko Ohnuki-Tierney, pp.206–35

p.202 **The Imperial Japanese Army**: *Kamikaze, Cherry Blossoms and Nationalisms* by Emiko Ohnuki-Tierney, p.110

p.202 **By the late nineteenth century**: 'Cherry Blossoms and Their Viewing' by Emiko Ohnuki-Tierney, in *The Culture of Japan as Seen Through Its Leisure*, edited by Sepp Linhart and Sabine Frustuck, Albany: State University of New York Press, 1998, p.223

p.203 **For soldiers poised for battle**: On 1 December 1941, a week before Japan attacked Pearl Harbor, a film called *Forty-Seven Rōnin* was released in Tokyo, directed by Kenji Mizoguchi. For tens of thousands of Japanese troops in Asia, the story reinforced the concepts of loyalty, self-sacrifice and honour that were central characteristics of the cherry ideology at the time.

p.203 **One man, Kiyoshi Hiraizumi**: *Nihonjin to Sakura* by Shōji Saitō, p.172. Saitō, a professor at Sōka University, wrote that the 'social cherry myth' was a deliberate creation by the state to achieve unified devotion to one person and one cause. The myth, he said, was disseminated during the 1930s and 1940s in repeated speeches and published papers by military and nationalistic scholars.

p.203 **'In case of emergency'**: *Nihonjin to Sakura* (*The Japanese and Cherry Blossoms*) by Shōji Saitō, Tokyo: Kōdan-sha, 1980, pp.127–8

p.203 **Hiraizumi's comments**: Takeshi Takagi, who taught Japanese classics at Nihon University, published a paper in 1938 called *Sakura and the Japanese Mentality*, which echoed Hiraizumi. The characteristics of cherry trees and the people's mentality were in total accord, he wrote, because the Japanese climate suited cherries perfectly, prompting flowers to bloom and drop en masse: 'The flowers start falling without any regrets while they are still gorgeously beautiful in colour and fragrance. The way they fall is like snow, creating a unique aesthetic beauty.' Takagi's paper is included in *Nihon Seishin to Nihon Bungaku* (*Japanese mentality and Japanese literature*), Fuzanbō, 1938.

34. *The* Somei-yoshino *Invasion*

p.204 **'a well-grown tree'**: *Ornamental Cherries*, p.217

p.204 **All that is known**: The origins of the *Somei-yoshino* are hotly debated. Early twentieth-century *sakura* authorities, such as Manabu Miyoshi and Gen-ichi Koizumi, interviewed many nurserymen in Somei village, but failed to discover who was first to develop and sell the variety. Thinking that *Somei-yoshino* must be related to Ōshima cherry, they went to Izu Peninsula and Ōshima island to look for it, but in vain. They also searched in the Yoshino mountains. Others theorised that the tree came from Saishu island in Korea, whilst the plant-hunter E.H. Wilson wrote that it was a hybrid of Ōshima cherry and Spring (*Edo-higan*) cherry. After the Second World War, Yo Takenaka of Keijo Imperial University flew to Seoul and debunked the Korean origin theory. He then planted *Somei-yoshino* seeds and, after observing the young trees' blossoms, concluded that Wilson was correct. After the war Takenaka became head of the Cytogenetics Department at the National Institute of Genetics. More recently, DNA analysis by researchers at the Shinrin Sōgō Kenkyū-jo in Tsukuba, Ibaraki Prefecture, has suggested that *Somei-yoshino* has genes from three wild cherries, not two: Ōshima cherry, Japanese mountain cherry and Spring cherry.

p.204 **Whereas *Yama-zakura***: *Sakura* by Toshio Katsuki, Tokyo: Iwanami Shinsho, 2015, pp.90–92

p.204 **All this meant**: In Tokyo's Ueno Park *Somei-yoshino* trees were first planted around 1876 and soon exceeded the number of *Yama-zakura* trees there. In 1883, 1,000 *Somei-yoshino* trees were planted in Mukōjima. In Asukayama 300 trees were planted in 1880, and 100 more in 1888. By 1892, 300 trees had already been planted at Yasukuni Shrine.

p.204 **Throughout Japan**: *Kamikaze, Cherry Blossoms and Nationalism* by Ohnuki-Tierney, p.121

p.205 **But by the beginning of the twentieth century**: *Saving the* Sakura by Akihito Hiratsuka, Tokyo: Bungeishunjū, 2001, pp.90–92

p.206 **'The famous old cherry tree locations'**: *Sakura* journal, 1936, no. 17

p.206 **'No flower has ever'**: Ibid.

35. *100 Million People, One Spirit*

p.207 **'There is earnestness in the rose'**: www.jjonz.us/RadioLogs, 11 April 1934

p.208 **Saitō was a popular diplomat**: After Saitō's death from cancer in 1939, Roosevelt arranged for his body to be returned to Japan aboard the USS *Astoria*. Ironically, the *Astoria* was sunk by the Japanese in August 1942 at the Battle of Savo Island.

p.209 **One dissenter was Yoshio Yamada**: Yamada's articles appear in a collection of his contributions called *Ō-shi* (*A History of Cherry Trees*).

p.209 **The 1940 issue**: *Sakura* journal, 1940, no. 21

36. *The Cherry and the* Kamikaze

p.212 **'Japan is in grave danger'**: *The Divine Wind: Japan's Kamikaze Force in World War II* by Rikihei Inoguchi and Tadashi Nakajima, Annapolis, MD: Naval Institute Press, 1958, p.19

p.213 **The *kamikaze* Special Attack Forces**: *Kamikaze* was the name of a typhoon in 1281 that destroyed an armada of ships under the command of Mongolia's Kublai Khan, which was set to invade Japan. *Kami* means god or deity, while *Kaze* means wind.

p.214 **'For the glory of the emperor'**: *Chiran Tokubetsu Kōgeki-tai*, compilation by Osamu Takaoka, Kagoshima: Japlan Ltd, 2009, p.78

37. Falling Blossoms

p.215 **Together with the 211**: For details of the conditions aboard USS *Whitehurst*, I am indebted to Max Crow, the yeoman and webmaster of the USS Whitehurst Association, at www.de634.org

p.216 **'The cherry blossoms have already'**: www.chiran-tokkou.jp/learn/pilots/, translation by Naoko Abe

p.216 **'The cherry blossoms are falling'**: *Konpaku no kiroku* (*A record of spirits*), compiled and published by the Chiran Tokkō Irei Kenshō-kai and Chiran Peace Museum management society. Kagoshima, 2004, p.111

p.217 **'I, Anazawa, no longer exist'**: Chiran Peace Museum collection

p.218 **'Anazawa's plane passed'**: *Chiran Tokubetsu Kōgeki-tai*, pp.48–9

p.218 **the so-called *Nadeshiko* girls**: *Nadeshiko* is a sweetly scented pink carnation, *Dianthus superbus*, which was treasured by Japanese aristocrats in feudal times. *Yamato Nadeshiko* means women of Yamato, symbolising the ideal of a beautiful, modest and gentle Japanese woman. The so-called *Nadeshiko* girls at Chiran helped the *kamikaze* pilots with chores, such as laundry and mending clothes, in the days before they flew.

p.219 **The explosion killed Irving Paul**: A memorial statue bearing the names of the men who died instantly when a *kamikaze* pilot flew into the port side of the USS *Whitehurst* is displayed at the Destroyer Escorts Historical Museum aboard the restored USS *Slater* in Albany, New York.

p.220 **In all, between October 1944**: The number of pilots who died and the number of Allied personnel who were killed vary considerably, depending on the source of the information.

38. Tome's Story

p.220 **'They did not hesitate'**: *The Mind of the Kamikaze* by Takeshi Kawamoto, Chiran: Peace Museum for Kamikaze Pilots, 3rd edition 2012, p.15

p.224 **'*Kamikaze* pilots are mere machines'**: An extremely well-read student, Ryōji Uehara had been deeply influenced by Benedetto Croce, an Italian philosopher who had opposed Mussolini's fascist movement. Near the secret testaments in the museum lay Uehara's favourite book, *Croce*, written by a Japanese historian, Gorō Hanyū, and published in 1939. On the back of the front cover, Uehara had written to his parents that he would be satisfied to die in the war, 'as I will have fought for Japan's liberty'.

Within the book itself, Uehara had encoded another secret: he was in love with a woman called Kyōko. He had met her when he was thirteen in a small town in the Nagano mountains. Kyōko's father was a soldier, and she became engaged to her father's subordinate and married him in August 1943.

'Look at the circled letters in the book,' Torihama said to me excitedly, pointing at *Croce*'s pages. Throughout the book Uehara had circled specific *kanji* and *hiragana* characters with a red pen. When stitched together, they read: 'Kyōko-chan, sayonara. I truly loved you. But you were already engaged and that was agony for me. I stopped myself telling you that I love you because I thought this would bring misfortune to you in the future. But I will always love you.' Kyōko died in the winter of 1943, struck down by tuberculosis. Uehara didn't hear the news until June 1944. 'I have also died today,' he wrote in his diary.

p.225 **'I will destroy the conceited'**: *Fading Victory: The Diary of Admiral Matome Ugaki, 1941–45*, translated by Masataka Chihaya, edited by Donald M. Goldstein and Katherine V. Dillon, Pittsburgh: University of Pittsburgh Press, 1991, p.610

PART SIX

39. *Children at War*

p.229 **four children and five grandchildren**: I am indebted to numerous members of the Ingram family for sharing details of their family history during interviews and email exchanges from 2014 to 2018.

40. *Black Christmas*

p.232 **Four more British nurses**: *Sisters in Arms* by Nicola Tyrer, London: Weidenfeld & Nicolson, 2008, p.67

p.232 **The events of that day**: Details of the Black Christmas incidents are based on Charles G. Roland's article, 'Massacre and Rape in Hong Kong', *Journal of Contemporary History*, January 1997, vol. 32, issue 1, also on testimony given at the International Military Tribunal for the Far East; on official Hong Kong, British and Canadian government records; and on Nicola Tyrer's *Sisters in Arms*. The Japanese army later tried to excuse the actions of its soldiers, saying that shots had been fired at Japanese troops within the hospital and 'it was impossible to tell if those lying in beds were indeed sick and wounded patients, or if they were in fact armed soldiers in disguise'; see 'Massacre and Rape in Hong Kong', p. 54. See also 'Judged War Crimes — British War Crimes Trial of Japan' by Hirofumi Hayashi, Tokyo: Iwanami Shoten, 2014, p.149.

p.232 **It wasn't until late 1942**: *Sisters in Arms* by Nicola Tyrer, pp.65–6

p.233 **The nurses did not have**: Ibid., pp.60, 62

p.233 **The POWs could walk**: Ibid., p.47

p.233 **One POW, Dominica Lancombe**: Interview with Dominica Lancombe, January 2018. After the war Dominica was taken to the Philippines, where she met her father, Alfred Taylor, an army nurse who had become a POW in Japan and worked down a mine after being captured.

41. *Protecting Benenden*

p.235 **'This sounds rather like a schoolboy trick'**: Ingram first outlined this strategy in a talk to his unit on 7 July 1940.

p.235 **Meanwhile, many single women**: *Benenden: A Pictorial History*, compiled by Michael Davies, CD edition, Benenden Parish Council, 2001, pp.67–8

p.235 **Three other German pilots**: In the 1960s, after the German government requested that their remains be returned home, the bodies were exhumed and given to the authorities. The man who both buried and exhumed the pilots was Sidney Lock, Ingram's longtime gardener. Interview with Anthony Price, son of Jessop Price, the vicar of St George's Church, Benenden, May 2015.

p.236 **It came as the weakening Nazi regime**: Germany later developed a more advanced missile, the V-2, which more successfully targeted London. One landed in Benenden on 28 January 1945.

42. *Ornamental Cherries*

p.240 **He described sixty-nine**: *Ornamental Cherries*, p.142
p.240 **'It permeates the air'**: Ibid., p.210
p.241 **'by far the most beautiful'**: Ibid., p.207
p.241 **'of a spoilt Victorian child'**: Ibid., pp.222, 224, 235

43. *Dark Shadows*

p.243 **Summers at The Grange**: Interview with Peter Ingram, May 2015. Interview with Heather Bowyer (née Ingram), May 2015.

p.244 **'Molly was a lot older'**: *Sisters in Arms* by Nicola Tyrer, p.67. After the war, Molly Gordon testified about the 'Black Christmas' atrocities at the International Military Tribunal for the Far East.

44. *Cherries of a 'Traitor'*

p.245 **Among the Allied powers**: From 'History Issues and the Path to Reconciliation – Recommendations to the Abe Administration' by Nobuko Kosuge; http://blogos.com/article/68027/. Statistics are from the International Military Tribunal for the Far East. Kosuge, a Professor of Law at Yamanashi Gakuin University, is an expert on British prisoner-of-war issues. The death rate of British soldiers held as POWs by the German army was about 5 per cent.

p.247 **In 1940, impressed by**: *Ōkashō* (*Excerpts on Cherry Blossoms*) by Tōemon Sano, Tokyo: Seibundō Shinkō-sha, 1970, p.55

p.247 **'to console the spirits'**: Sano gave 2,000 more cherries to the Maizuru Naval base in Kyoto in 1942. And the navy commissioned him to deliver a further 2,000 cherries to Shanghai in March 1943, mostly *Yama-zakura* and some flowering varieties, such as *Kanzan* and *Fugenzō*. Ibid., p.55

p.248 **'Count Ōtani's great ambitions'**: Ibid., p.57

p.248 **'I spent several days chopping'**: Ibid., p.68

p.248 **'I believed in Japan's victory'**: Ibid., p.69

45. Britain's Cherry Boom

p.249 **'If readers of Captain Ingram's book'**: 'The Blossom on the Bough' by Victoria Sackville-West, *The Observer* newspaper, 1948

p.250 **'excellently conceived'**: 'Ornamental Cherries: A review' by Walter B. Clarke, *American Nurseryman*, 15 August 1948

p.251 **'"Cherry" influenced my life'**: Interviews with Lady Anne Berry in December 2015

p.261 **Sixty years later**: Phone interview with Jonathan Webster, November 2015

46. Ingram's 'Royal' Cherries

p.255 **In July 1948 Sir Eric**: The grafting of scions to root stock was usually carried out in early spring. In this case, however, Sir Eric attached grafts that already bore shoots to the bark of the stock wood in July. I am grateful to Mark Flanagan, the Crown Estate's Keeper of the Gardens, for unearthing letters in the estate's archives sent to and from Savill and Ingram. Flanagan died in October 2015.

p.255 **Sir Eric asked Ingram / Ingram replied**: Correspondence between Sir Eric Savill and Collingwood Ingram, 13 April, 15 April and 19 April 1949. Courtesy of the Crown Estate archives.

47. The Somei-yoshino Renaissance

p.257 **'It was as if the clocks'**: *Saving the* Sakura by Akihito Hiratsuka, Tokyo: Bungeishunjū, 2001, pp.158–9

p.258 **'Somei-yoshino bubble'**: Ibid., p.159

p.258 **By 1964**: Ibid.

p.258 **By the beginning of**: Ibid., p.189

p.259 **Few of them grew**: One exception is Shinjuku Gyoen National Garden in central Tokyo. Once the site of a *daimyō* residence, it was later managed by the Imperial Household Agency. It was opened to the public after the war and now features more than 1,100 flowering trees of sixty-five different varieties. They include Collingwood Ingram's *Okame* cultivar, *Taihaku* and numerous cherries that once grew on the Arakawa riverbanks, such as *Amanogawa, Arashiyama, Chōshū-hizakura, Shirotae, Shōgetsu* and *Surugadai-nioi.*

p.260 **'They developed an image'**: Interview with Tōru Koyama, Tokyo, December 2014

p.261 **One vocal critic**: Interview with the sixteenth Tōemon Sano, Kyoto, December 2014

PART SEVEN

48. *A Garden of Memories*

p.266 **'That a people who'**: *A Garden of Memories*, p.70

p.266 **He was particularly concerned**: *Random Thoughts*, pp.6–7

p.267 **Among these were two flowers**: *Garden of Memories*, pp.79–80. *Rubus* 'Benenden' was a cross between a plant from an extinct volcano in Mexico and the *Rubus deliciousus*, also known as the Rocky Mountain raspberry. In the centre of the hybrid shrub's white blooms was a crown of golden anthers. Ingram exhibited the plant at the RHS in 1947. 'Benenden Blue' was a rosemary plant with sapphire-blue flowers that Collingwood had found by chance in the 1930s in south-west Corsica. In the mountains near Sartene, Ingram's chauffeur-driven car collided with a coach. As the two drivers exchanged insults, Ingram strolled off to look for plants. On top of a small knoll he noticed the feathery-leaved shrub and gathered some seeds. Within weeks, they were sprouting at The Grange.

p.267 **'He often told me'**: Interview with Moira Miller in November 2014

p.267 **Even in his eighties and nineties**: Interview with Sibylle Kreutsberger, May 2015

p.268 **In his later years**: Ingram's favourite rhododendrons included the pale-pink 'Infanta', which won an RHS Award of Garden Merit in 1941, and the red 'Timoshenko', both of which he gave to Lady Anne Berry for Rosemoor Garden. Other still-popular Ingram crosses included the 'Sarled', 'Muy Lindo', 'Flamenco', 'Carolyn Hardy' and 'Captain Blood' varieties.

p.269   **Smith reported that**: Interview with Lawrence Smith, February 2015

p.270   **'coincide with those of'**: *Random Thoughts*, preface

p.271   **'Mr Ingram was very famous'**: Interview with Roland Jefferson, January 2015. Jefferson moved to Hawaii with his Japanese wife after retirement.

49. A Peaceful Death

p.272   **'They were very different characters'**: Interview with Ruth Tolhurst, October 2014

p.273   **'cavalcade of famous plant hunters'**: 'From the Golden Age' by Roy Lancaster, *GC & HTJ* journal, 31 October 1980, pp.22–3

p.274   **'Plants, which once seemed plentiful'**: 'For the Record' by Michael Zander, *The Garden*, April 1980, p.159

p.275   **'Days before his death'**: Interview with Charlotte Molesworth, October 2014

50. The Grange after Ingram

p.276   **'indefatigable traveller'**: *The Times*, 22 May 1981, p.16

p.276   **'"Cherry" was the last'**: Interview with Peter Kellett, October 2014

p.277   **After Ingram's death**: Martin Miller arranged for three flower experts – William Nelmes, Jr, Tom Wright and Alan Hardy – to visit The Grange and identify rare specimens. In a letter to Miller on 12 October 1983, Nelmes, an adviser to the RHS after his retirement as director of Cardiff Parks, noted that some trees had been affected with honey fungus and suggested that mulches of peat, bark chippings or leaf mould be used to smother weeds, conserve moisture and improve the soil.

p.277   **Parsons, a recording engineer**: Alan Parsons interview with Kent Barker, *Benenden Magazine*, October 2017, pp.18–19

p.278   **They included a rare**: Kingdon-Ward had discovered the tree at 7,000 feet in northern Burma. *Ornamental Cherries*, p.115

p.278   **Despite the destruction**: Letter from Carolyn Hardy, county organiser of the National Garden Scheme, to Charlotte

Molesworth, 12 July 1988. The event raised £369. Hardy called it 'a very courageous decision' to open the garden and noted that the Parsonses had been 'caring custodians' of this 'extra special' garden. The Parsonses decided to sell The Grange that spring.

p.278 **Soon after the storm**: The Parsonses sold The Grange to Anthony Biddle, who took down the recording studio. Biddle, in turn, sold the property in 1992 to a businesswoman, Linda Fennell, who transformed it into a live-in facility for people with learning difficulties. Fennell sold the property in 2016 to a businessman called Andrew Scott.

51. *Home and Abroad*

p.278–9 **Britain cannot get enough**: Interview with Nick Dunn, February 2018

p.279 **Tony Kirkham**: Interview with Tony Kirkham, January 2015

p.280 **'When the pure-white blossoms'**: Interview with Jane Percy, 12th Duchess of Northumberland, May 2017

p.281 **'These cherries are a sign'**: Interview with Canon Paul Scott, May 2017

p.282 **And in another ambitious initiative**: After Japanese Prime Minister Shinzō Abe met British Prime Minister Theresa May in August 2017, their joint declaration included a proposal from the Japan Association, led by Sandy K. Sano, to donate 'a substantial number of cherry blossom trees to the UK'. The first trees will be planted in London's Regent's Park, Greenwich Park, Richmond Park and Bushy Park.

52. *The Next Generation of* Sakuramori

p.284 **As they walked to and from**: Speech by Frank Planton on 4 June 2006, at Hakodate YMCA. Captured in Java, Planton arrived in Hakodate in November 1942 and worked at the dock, cleaning and painting ships.

p.284 **Many of the prisoners**: More than 1,500 POWs remained in the Hakodate camps at the end of the war. POW Research Network Japan (co-founders: Aiko Utsumi and Tōru Fuku-bayashi), online document at http://www.powresearch.jp

p.285 **'notorious hell-hole'**: 'A Model Japanese' by Peter V. Russo, *The Argus* (Melbourne, Victoria), 15 October 1948

p.285 **Conditions in the camp**: On 24 March 1948 Colonel Toshio Hatakeyama was sentenced to twelve years' imprisonment at the International Military Tribunal for the Far East at Yokohama, for the mistreatment of Hakodate prisoners under his command between December 1942 and March 1944. The POWs considered Hatakeyama's successor, Lieutenant Colonel Shigeo Emoto, a fluent English speaker, to be compassionate and fair. However, Emoto was criticised for 'interpreting the samurai code in a manner detrimental to Japan' and was removed as camp commandant. After the war, several POWs visited him and his son, Susumu, to thank him for his kindness.

p.286 **'lack of pride and shame'**: 'POWs Awake to Our *Bushidō*', *Hokkaido Shimbun*, 6 March 1943. 'The Psychology of the American-British POWs That Japanese Can't Understand', *Hokkaido Shimbun*, 17 March 1943

p.286 **'All the botanists'**: Interview with Masatoshi Asari, February 2018

p.288 **'One rare variety'**: Interviews with Chris Sanders in October 2014, May 2015 and February 2018

p.288 **Joining Sanders on many**: Interviews with Chris Lane, November 2014 and February 2018. Lane is the holder of five National Plant Collections: flowering cherries, hamamelis, wisteria, amelanchier and parrotia.

p.288 **As their friendship grew**: Lane and Sanders had most success in Belgium, particularly at the Pépinière Choteau nursery east of Mons, and at the 160-year-old Arboretum Kalmthout and the Hemelrijk estate near Antwerp. The latter two were owned by husband and wife Robert and Jelena de Belder and Robert's elder brother, Georges, all passionate horticulturalists who had known Collingwood Ingram.

p.289 **After referring to**: Published by the Flower Association of Japan in 1983, the translation of the *Manual of Japanese Flowering Cherries* was one of the first books in English listing all the Japanese flowering-cherry varieties.

p.290 **'With vivid memories'**: *New York Times*, 6 October 1971, p.1

53. Cherries of Reconciliation

p.292 **These cherries now grow**: Some *Matsumae* varieties have been given trade names to make them more marketable. For example, *Matsumae-fuki*'s trade name is *Chocolate Ice* and *Matsumae-shizuka* is sold as *Fragrant Cloud*.

p.292 **And the royal family**: Interview with Harvey Stephens, deputy keeper of the gardens at Windsor Great Park, February 2018

p.293 **'As the *Matsumae* cherries'**: Interview with Quentin Stark, February 2018

p.293 **And so on Thursday, 2 March 2000**: The Grange is on a list of historic parks and gardens in the borough of Tunbridge Wells.

p.294 **'It's miraculous'**: Interview with Masatoshi Asari, February 2018

EPILOGUE

54. Millennia Trees

p.301 **Maeda and a team**: In 1985 the local government built a cherry museum next to the tree to explain Maeda's grafting techniques. The museum also highlights the efforts of Chiyo Uno, a best-selling female author, to save the tree in the late 1960s.

p.302 **To protect the *Usuzumi-zakura***: Interview with Takashi Sanbongi, leader of the volunteer group, March 2017

55. The Great Wall of Cherry Blossoms

p.304 **'We were told the nuclear plant'**: Interview with Tadashige Shiga, February 2018

p.304 **'The cherries we're planting now'**: Ibid.

p.305 **'If we plant 99,000 trees'**: Interview with Cai Guo-Qiang, NHK television, 14 January 2013

p.305 **What varieties survived**: In August 1910 a series of unlikely events began that were to ensure the Arakawa cherries' survival to the present day. They started with a destructive storm that poured torrents of water into the Arakawa River, which burst its banks, submerging the cherries. The Tokyo municipal government started building a drainage canal to divert future flood waters. Concerned that the cherry varieties would disappear, Funatsu sent scions from the varieties to a distant relative, Dentarō Matsumoto, who owned a nursery in Shingō village, north of Tokyo. After grafting Funatsu's scions, Matsumoto planted them behind his house. Matsumoto later asked a plant grower in a village near Shingō called Angyō to take the young trees. That grower, Kamenosuke Koshimizu, kept the cherries in his nursery for more than a decade. On the riverbank itself, the number of different varieties had fallen from seventy-eight in 1886 to thirty-two in the 1930s. During the Pacific War the military authorities ordered Koshimizu to cut down the trees and use the land to grow food. He refused and paid off the officials.

One of the few people who knew where Funatsu had sent scions was his grandson, Kanematsu, who became a cherry specialist himself, studying plant genetics with a Tokyo University expert called Yoshito Shinotō. After Shinotō took a job in 1949 at the National Institute of Genetics, he and Kanematsu persuaded Koshimizu to move the Arakawa cherries to the facility. Finally in the 1960s, the government set up a conservation forest, called Tama Forest Science Garden, west of Tokyo. All the descendants of the Arakawa cherries at the genetics institute were moved to that forest, together with descendants of historic or renowned cherry trees from throughout Japan. They are now part of a collection of more than 1,300 cherry trees.

p.306 **A few, including**: *Arakawa no Goshiki Zakura* (*Five Coloured Cherries of Arakawa*) by Keiichi Higuchi, Tokyo Agriculture University, 2013

p.307 **Yet since 2010**: Interview with Tetsu Tada, curator at the Koganei Cultural Property Centre, December 2017

p.307 **Within a decade**: In central Tokyo twenty-six *Okame* cherry trees, first created at The Grange by Ingram, were planted along a street in the Nihonbashi district in 2006.

p.308 **In my head**: 'A very thin veil of conservatism covers Japan these days,' Osamu Aoki, a well-known political commentator, told me in an interview in March 2017.

p.308 **And he railed against**: Nearly half of all magnolia species in the wild are threatened with extinction, as are a quarter of rhododendron species and a third of maple species, according to 'Integrated conservation of tree species by botanic gardens: a reference manual', compiled by Sara Oldfield and Adrian C. Newton, Richmond, UK: Botanic Gardens Conservation International, November 2012

p.308 **'Progress, improvement'**: From a 1926 Collingwood Ingram journal entry, published in *Isles of the Seven Seas*, p.133

Bibliography

Aung San Suu Kyi, *Letters from Burma*, London: Penguin Books, 1997

Bailey, Isabel, *Herbert Ingram, Esq., M.P.*, Boston: Richard Kay, 1996

Benesch, Oleg, *Inventing the Way of the Samurai*, Oxford: Oxford University Press, 2014

Chamberlain, Basil Hall, *Things Japanese*, London: John Murray, 5th revised edition, 1905

Coats, Alice M., *The Quest for Plants: History of the Horticultural Explorers*, Littlehampton Book Services, 1969

Davies, Michael (comp.), *Benenden: A Pictorial History*, CD edition, Benenden Parish Council, 2001

Fairchild, David, *The World Was My Garden: Travels of a Plant Explorer*, New York: Charles Scribner's Sons, 1938

Farrer, Reginald J., *The Garden of Asia — Impressions from Japan*, London: Methuen & Co., 1904

Fortune, Robert, *Yedo and Peking*, London: John Murray, 1863

Freeman-Mitford, Algernon, *Tales of Old Japan*, London: Macmillan and Co., 11th edition, 1910

Goffin, Magdalen, *The Watkin Path. An Approach to Belief*, Brighton: Sussex Academic Press, 2006

Hayashi, Hirofumi, *Sabakareta Senso Hanzai (Judged War Crimes)*, Tokyo: Iwanami Shoten, 2014

Hearn, Lafcadio, *Glimpses of an Unfamiliar Japan*, Boston: Houghton Mifflin, 1894

Higuchi, Keiichi, *Arakawa no Goshiki-zakura (The Goshiki-zakura of Arakawa)*, Tokyo: Tokyo University of Agriculture Press, 2013

Hiratsuka, Akihito, *Sakura o Sukue (Saving the* Sakura*)*, Tokyo: Bungei-shunjū, 2001

Holmes, Keiko, *Agape: A journey of Healing and Reconciliation*, Tokyo: Inochi no Kotoba-sha, 2003

Ingram, Collingwood, *Isles of the Seven Seas*, London: Hutchinson & Co., 1936

—— *Ornamental Cherries*, London: Country Life, 1948

—— *A Garden of Memories*, London: H.F. & G. Witherby Ltd, 1970

—— *The Migration of the Swallow*, London: H.F. & G. Witherby Ltd, 1974

—— *Random Thoughts on Bird Life*, self-published, 1978

Inoguchi, Rikihei and Nakajima, Tadashi, *The Divine Wind: Japan's Kamikaze Force in World War II*, Annapolis: Naval Institute Press, 1958

Kaempfer, Engelbert, *The History of Japan*, published posthumously in 1727; Glasgow: James MacLehose and Sons, 1906 edition

Katō, Yōko, *Sensō no nihon kin/gendai-shi* (*Japanese Modern and Contemporary History of War*), Tokyo: Kōdan-sha, 2002

—— *Manshū Jihen kara Nicchū Sensō e* (*From the Manchurian Incident to the Second Sino-Japanese War*), Tokyo: Iwanami Shoten, 2007

—— *Soredemo Nihonjin wa 'Sensō' o eranda* (*The Japanese Still Chose the 'War'*), Tokyo: Shinchō-sha, 2016

Katsuki, Toshio, *Nihon no sakura* (*The Japanese Cherries*), Tokyo: Gakken Kyōiku Shuppan, 2014

—— *Sakura*, Tokyo: Iwanami Shoten, 2015

—— *Sakura no kagaku* (*Science of Sakura*), Tokyo: SB Creative, 2018

Kawamoto, Takeshi, *The Mind of the Kamikaze*, Chiran: Peace Museum for Kamikaze Pilots, 2012

Kihara, Hiroshi, Tanaka, Hideaki, Kawasaki, Tetsuya and Oba, Hideaki, *Shin Nihon no Sakura*, Tokyo: Yama-kei Publishers, 2007

Kōhoku History Association, *Kōhoku no Goshiki-zakura*, Tokyo: Kōhoku no Rekishi o Tsutaeru Kai, 2008 and 2015

Kosuge, Nobuko Margaret, *Poppy to Sakura* (*Poppies and Cherry-blossoms*), Tokyo: Iwanami Shoten, 2008

Kuitert, Wybe, *Japanese Flowering Cherries*, Portland, Oregon: Timber Press, 1999

Linhard, Sepp and Frustuck, Sabine (eds), *Cherry Blossoms and Their Viewing*, Albany: State University of New York Press, 1998

McClelland, Ann, *The Cherry Blossom Festival: Sakura Celebration*, Boston: Bunker Hill Publishing, 2005

Mitani, Taichirō, *Nihon no Kindai towa Nan de attaka* (*What were the Japanese modern times?*), Tokyo: Iwanami Shoten, 2017

Mizukami, Tsutomu, *Sakuramori*, Tokyo: Shinchō-sha, 1976

Murakami, Shigeyoshi, *Kokka Shinto* (*State Shinto*), Tokyo: Iwanami Shoten, 1970

Nitobe, Inazō, *Bushidō: The Soul of Japan*, Tokyo: Teibi Publishing Co., 13th edition, 1908, Project Gutenberg e-book

—— *Bushidō in Contemporary Language*, Tokyo: Chikuma Shobō, 2010

North, Marianne, *Recollections of a Happy Life, Being the Autobiography of Marianne North*, ed. Janet Catherine Symonds, New York: Macmillan & Co., 1894

Ogawa, Kazusuke, *A Literary History of Cherries*, Tokyo: Bungeishunjū, 2004

Ohnuki-Tierney, Emiko, *Kamikaze, Cherry Blossoms and Nationalism: The Militarization of Aesthetics in Japanese History*, Chicago: University of Chicago Press, 2002

—— *Nejimagerareta Sakura* (*Distorted Cherry Blossoms*), Tokyo: Iwanami Shoten, 2003

—— *Flowers That Kill: Communicative Opacity in Political Spaces*, Redwood City: Stanford University Press, 2015

Okuda, Minoru, Kihara, Hiroshi and Kawasaki, Tetsuya, *Nihon no Sakura*, Tokyo: Yama-kei Publishers, 1993

120 Years at the Imperial Hotel, Tokyo: Imperial Hotel Ltd, 2010

Pollard, Ernest and Strouts, Hazel (eds), *Wings Over the Western Front: The First World War Diaries of Collingwood Ingram*, Charlbury, Oxfordshire: Day Books, 2014

Robinson, William, *The English Flower Garden*, 1900 edition

Roland, Charles G., 'Massacre and Rape in Hong Kong', *Journal of Contemporary History*, 32 (1), January 1997

Ruoff, Kenneth J., *The People's Emperor*, Cambridge, MA: Harvard University Asia Center, 2002

Saitō, Shōji, *Nihonjin to Sakura* (*The Japanese and Cherry Blossoms*), Tokyo: Kōdan-sha, 1980

Sakura journal, Journal of the Cherry Association 1918–1943, Shōwa edition, Tokyo: Ariake Shobō, 1981

Sano, Tōemon, *Ōkashō* (*Excerpts on Cherry Blossoms*), Tokyo: Seibundō Shinkō-sha, 1970

Sargent, Charles Sprague, *Forest Flora of Japan*, Boston: Houghton, Mifflin & Co., 1894

Sasabe, Shintaro, *Sakura Otoko Gyōjō*, Tokyo: Heibon-sha, 1958

Satō, Toshiki, *Sakura ga Tsukutta 'Nihon'* (*The Japan that the Cherry Blossoms Made*), Tokyo: Iwanami Shinsho, 2005

Shimazono, Susumu, *Kokka Shinto to Nihonjin* (*State Shinto and the Japanese*), Tokyo: Iwanami Shoten, 2010

Shinrin Sōgō Kenkyū-jo, *Sakura Hozon-rin Guide* (*A Guide to the Sakura Reserved Forest*), Tokyo: Shinrin Sōgō Kenkyū-jo, 2014

Shirahata, Yōzaburō, *Hanami to Sakura*, Tokyo: PHP Research Institute, 2000

Suzuki, Yoshikazu, *Sakuramori Sandai* (*The Three Generations of Sakuramori*), Tokyo: Heibon-sha, 2012

Takagi, Toshiro, *Tokkō Kichi Chiran*, Tokyo: Kadokawa, 1973

Takaoka, Osamu (comp.), *Chiran Tokubetsu Kōgeki-tai*, New Edition, Kagoshima: Japlan Ltd, 2009; Tokyo: Heibon-sha, 2012

Takeda, Kiyoko, *Tennō kan no Sōkoku* (*Conflict of Views on the Emperor*), Tokyo: Iwanami Shoten, 1976

Torigoe, Hiroyuki, *Hana o Tazunete Yoshinoyama*, Tokyo: Shūei-sha, 2003

Torihama, Akihisa, *Chiran: Inochi no Monogatari* (*Chiran: A Story of Life*), Tokyo: Kizuna Shuppan, 2015

Tyrer, Nicola, *Sisters in Arms*, London: Weidenfeld & Nicolson, 2008

Uehara, Ryōji, *Ah Sokoku yo Koibito yo*, Nagano: Shinano Mainichi Shimbun-sha, 2005

Ugaki, Matome, *Fading Victory: The Diary of Admiral Matome Ugaki, 1941–45*, trans. Masataka Chihaya, ed. Donald M. Goldstein and Katherine V. Dillon, Pittsburgh: University of Pittsburgh Press, 1991

Yamada, Yoshio, *Ō-shi* (*History of Cherry Blossoms*), Tokyo: Kōdan-sha Gakujutsu Bunko, 1990

Websites

Nippon Yūsen: https://www.nyk.com/ir/investors/history/

Pollard, Ernest, *History of Benenden*: http://www.benenden.history.pollardweb.com/

POW Research Network Japan: http://www.powresearch.jp/

Shibusawa Eiichi Memorial Foundation: https://www.shibusawa.or.jp/

Journals

I have referred to numerous articles written by Collingwood Ingram in the years 1923–59 in British horticultural magazines, including *The Garden*, *The Gardeners' Chronicle*, *Gardening Illustrated* and *Journal of the Royal Horticultural Society*, as well as articles in the *Illustrated London News*. I have also referred to several magazine and newspaper articles written about Ingram.

Acknowledgements

When friends first suggested that I should write an adaptation of the *'Cherry' Ingram* book in English, the idea seemed implausible. I had written a few magazine articles in English but had never contemplated a book. This one materialised only because I received the support and encouragement of countless people.

As with the original Japanese-language edition, I am deeply indebted to Ernest Pollard for his cooperation, wisdom and friendship. Not only did Ernie provide me with Ingram's diaries, sketches and photos for the Japanese version, he offered further materials for the international adaptation. Ernie was generous to a fault in answering questions and reading the manuscript. There is no greater authority on Collingwood Ingram.

For the English-language edition, I was extremely fortunate to meet a caring literary agent in London, Patrick Walsh of PEW Literary, with whom I felt an immediate rapport. Patrick guided me through the challenge of writing an English book in a foreign country, enabling me to overcome my initial anxieties. In turn, he introduced me to the charming publishing director of Chatto & Windus, Clara Farmer, and she assigned a patient and talented editor, Charlotte Humphery, to work with me. This trio – Patrick, Clara and Charlotte – made *'Cherry' Ingram* a reality. I feel lucky and blessed to have become connected to such a dedicated, experienced and friendly network of people in the English literary world. John Ash and Margaret Halton, the staff at PEW Literary, also supported me. To move the project forward, a team of graduate students from Leeds University's East Asian Studies department first translated the Japanese-language edition into English. The students, led by Soeren Otter-Sharp, were Yasuko Arakawa, Finn Catterall, Yu-Jou Chen, Nicole Churchill, John Lowe, Gillian Melton, Nick E. Ruban and Elizabeth Tiu. Their translation became the basis for this adaptation.

For help with writing, I turned to my journalist husband, Paul Addison. Every weekend and on countless evenings for almost two years Paul and I discussed the ingredients that English-language readers would find illuminating about the Ingram family, the history of cherries and UK–Japan relations.

We went back and forth on the text. I wrote and he polished and edited my English while suggesting new angles and perspectives. In that sense, we shared the journey together, and I regard this adaptation as a joint effort. Nonetheless, I take full responsibility for the book's contents and any errors.

Multiple other people deserve my thanks for helping bring 'Cherry' Ingram to life. First, I am grateful to the people I interviewed over four years, including many of Ingram's relatives and descendants. In all cases, they gave their time and knowledge unselfishly. In particular, I am thankful to the UK's two leading cherry experts, Chris Sanders and Chris Lane, for their unstinting advice about cherry trees. Chris Sanders read the manuscript and corrected errors on cherry nomenclature and biology.

I am also truly appreciative of the contributions made by my friends David McAlpine, Angelina Skelton, Alpheus Boileau and Graham Hillier for reading the raw text and offering thoughtful opinions. Thanks also to Professor Mark Williams, a UK specialist in Japanese studies, for reading and checking the Japanese history even after he left Leeds University to become a vice president at my alma mater, the International Christian University, in Tokyo.

At Chatto & Windus, I am also indebted to Lily Richards and Stephen Parker for their work on the cover of this book. I would also like to praise copy-editor Mandy Greenfield, proofreader Anthony Hippisley and indexer Vicki Robinson, who read and corrected the manuscript during the final stages of publication.

A couple of points about the style. First, I decided to write Japanese people's names the Western way, with their surname after the given name, i.e. Naoko Abe. In Japan, the surname comes first so I am known there as Abe Naoko. A few internationally recognised names such as Katsushika Hokusai or Utagawa Hiroshige are written in Japanese style with the surnames first. Second, the names of cherry varieties in Japanese are written in italic such as *Taihaku* and *Hokusai* in accordance with the style of the text.

My final thanks are to my family: my parents, Hiroyoshi and Akiko Abe, to whom I owe so much for their lifelong support; to my sister Junko, and to my two sons, Sean and Kenji, for their unwavering encouragement during the writing of this book.

Index